Nematode Management in Plants

Nematode Management in Plants

– *Author* –
Mukul Tomar

Wordpen Academics
Publishing Science That Matter

ISBN: 978-93-90438-95-2 (Hardback)

Published by : **BIO-GREEN BOOKS**
4736/23, Ansari Road, Darya Ganj,
New Delhi – 110 002
Phone: 011-23245578, 23244987
E-mail: biogreenbooks@gmail.com

Printed at : **Replika Press Pvt. Ltd.**

Preface

Nematodes are the most numerous multi-cellular animals on earth. A handful of soil will contain thousands of the microscopic worms, many of them parasites of insects, plants or animals. Free-living species are abundant, including nematodes that feed on bacteria, fungi, and other nematodes, yet the vast majority of species encountered are poorly understood biologically. There are nearly 20,000 described species classified in the phylum Nemata. Nematodes are structurally simple organisms. Adult nematodes are comprised of approximately 1,000 somatic cells, and potentially hundreds of cells associated with the reproductive system . Nematodes have been characterized as a *tube within a tube*; referring to the alimentary canal which extends from the mouth on the anterior end, to the anus located near the tail. Nematodes possess digestive, nervous, excretory, and reproductive systems, but lack a discrete circulatory or respiratory system. In size they range from 0.3 mm to over 8 meters. Some species of roundworm may contain more than 27 million eggs at one time and lay more than 200,000 of them in a single day. Some scientists have estimated that there may be as many as half a million more unknown species of roundworm yet to be discovered, an estimate based on the fact that many new species are still being discovered, that relatively few people are looking for more species, and that most roundworms look pretty much alike. If the estimated number of species is anywhere close to correct, it would mean that roundworms are the second most diverse group of animals, trailing behind only the arthropods.

Nematodes were once classified with a very large and heterogeneous cluster of animals grouped together on the basis of their overall worm-like appearance, simple structure of an internal body cavity called a pseudocoelom, and the lack of features such as cilia and a well-defined head that are found in most animals. This group, variously known as Aschelminths or Pseudocoelomata, is today no longer recognized as a natural one. It is quite likely that the simple body plan of these organisms has resulted from reduction and simplification from more than one group of ancestral organisms, and so the pseudocoelom is neither a uniquely derived nor useful character. The simplicity is thus a result of secondary simplification from a more complex body design, and not necessarily an indication of primitive or simple origins. Current studies indicate that nematodes are actually related to the arthropods and priapulids in a newly recognized group, the Ecdysozoa. Nematodes feed on all parts of the plant, including roots, stems, leaves, flowers and seeds. Nematodes feed from plants in a variety of ways, but all use a specialized spear called a stylet. Note the differences in stylet length and shape. The size and shape of the stylet is used to classify nematodes and also can be used to infer their mode of feeding. There are several methods commonly used to control plant-parasitic nematodes. These methods can be divided in to three main types: biological control, cultural control and chemical control. The most practical form of biological control is the use of nematode-resistant plants. In this control method, plant breeders cross natural nematode resistance genes into cultivated plant species to improve their resistance to nematodes.

Nematodes represent a unique challenge to agricultural research, in that they combine the potential for serious reductions in growth and yield in a wide range of crop plants, often with rather nonspecific and easily misdiagnosed symptoms. Development of the concept of pest management and their implementation have led to a greater appreciation of the need for a wide range of tactics for nematode control.

The present book "Nematode Management in Plants" provides an authoritative review of many aspects of nematode control and progress in the field of nematode management programme. This book will be useful to plant pathologist, nematologist, research and extension workers, teachers and students.

— Dr. Mukul Tomar

Contents

Chapter 1

Introduction

Plant parasitic nematode, the hidden enemy of crops is one of the many groups of harmful organism which depend on plants for their survival. Nematode can cause damage to almost all kinds of crops, however, due to their subterranean habit, microscopic size (from 0.3 to 10 mm length), they are invisible to the naked eye. They penetrate and feed on the root of growing plants, stealing nutrients vital for plant growth and exposing the roots to attack by other soil pathogens. It has widely been recognized that plant parasitic nematodes constitute one of the most devastating pests groups and are responsible for insidious disease symptoms in different crops causing huge losses. Estimated annual yield losses in the world's major crops due to plant parasitic nematodes is about 12.3% and it is about 14% in the developing countries. In India, recent estimate showed nematode is responsible for both quantitatively and qualitatively yield losses amounting about Rs.240 billion every year. Beside direct damage, plant parasitic nematodes serve as predisposing agents in development of disease complexes with fungi, bacteria and viruses. In many situations, plant varieties resistant to fungi, bacteria are rendered susceptible when parasitized by nematodes. There is no doubt that nematodes either alone or in combination with other pathogens constitute an important constraint to world food production. Intensive and extensive cultivation of crops particularly in irrigated crop production system has seriously aggravated nematode problems in various crops. The hidden nature of nematode causing damage out of sight of farmers, scientists and non specific disease symptoms in the above ground parts of the crops are perhaps main reasons why so little attention has been given to the hidden pest of crops.

Management of plant parasitic nematodes with the use high doses of DD, EDB, DBCP etc. had been found promising, though did not receive much popularity. Moreover, all the effective chemicals have been withdrawn from world market due to their harmful effects on environment. Subsequently, efforts have been made to search for newer chemicals among the group of non-fumigants but failed to achieve the effective control as that of fumigants. In fact, nematodes are comparatively hardy animals require high doses of insecticide having nematicidal property. The growers are still dependent on the limited number of insecticides only because of non-availability of true and effective nematicides. With the increasing concern on environment, various alternative pest control methods like cultural, physical and biological control methods and botanicals are being tried to reduce the nematode damage of crops. However, judicious use of chemical nematicides could be applied for protection of many crops. Integration of various available practices is one of the current approaches for managing pest problems of crops. Cultural practices are known from time immemorial as a multiple pest control strategy. Biopesticides of botanical origin have also been proved as effective alternative of nematicides. Therefore, current options for nematode management are cultural practices, physical methods, biointensive nematode suppression, botanicals and sensible use of chemical nematicides.

Considering importance of nematodes in the integrated pest management system, following low input plant protection technologies viz. summer ploughing, soil solarization, organic manuring, crop rotation, adjusting dates of sowing, growing resistant varieties, irrigation management, optimum fertilization, hot water treatment, clean cultivation, green manuring, inter/mixed cropping, judicious use of pesticides, integration of two or more above mentioned methods could be adopted for managing insect pest, diseases including nematodes in the crop production system.

Nature of Nematode Problems

Plant parasitic nematodes can be detrimental to crop growth and development depending on population density and host susceptibility. Generally, they feed on the host tissues with the help of their protrusible stylet causing plants injury and due to feeding and secretion modify the host tissue into specialized nutritive cells as multinucleate giant cell, syncytium or nurse cells for ensuring permanent feeding. Some other nematodes induce gall formation on plant's root, leaf and seed. While feeding on plant tissues, develop lesion as a result of cell death and subsequent discolourations. Infected plants are easily attacked by various soil pathogens like bacteria, fungi and develop a disease complex/diseases syndrome. The etiology of those diseases caused by the organisms involved is difficult to determine. Several nematodes serve as vector of plant viruses. Thus nematode functions as plant pathogen, predisposing agent and vector of plant viruses. Plant parasitic nematodes are known to interfere the activity of beneficial nitrogen fixing Rhizobium bacteria in leguminous crops.

Root Knot Nematodes (*Meloidogyne* Spp.)

Root-knot nematodes (Meloidogyne spp.) are global menace to crop production. It has a very wide distribution and causes serious damage to crops particularly in vegetables. The average yield losses in the world are believed to be about 5% and could be more in the developing countries of tropic and sub-tropic. Considering the universal significance of root-knot nematodes, an International Meloidogyne Project (IMP) was operated (1975-1984) with its head quarter at North Carolina State University, USA and its collaborating centres were in many developing countries of the tropics and subtropics. Worldwide more than 97 known species of root-knot nematode have been recorded and only 14 known species of Meloidogyne are recorded in India. Four species of root knot nematodes viz. Meloidogyne indica, M. lucknowica, M. triticooryzae and M. piperi have been described from India. Various insect-pests, diseases and weeds are inflicting damages to vegetable crops. Root-knot nematodes (Meloidogyne spp.) are one of the potential constraints for cultivation of vegetables particularly in the developing countries of tropics and subtropics. Vegetable crops harbour large number of plant parasitic nematodes but root-knot nematode is most damaging one. It affects the crop directly and indirectly by interaction with various soil borne fungi, bacteria and viruses.

The most predominant species of root-knot nematodes are Meloidogyne incognita, M. javanica, M. arenaria and M. hapla. All the species of root-knot nematodes produce a characteristic 'root gall 'or'knotted root symptom, which could be easily recognized by naked eye. There is hardly any vegetable crop which is not attacked by the root-knot nematodes. Therefore, it has widely been considered as a limiting factor for cultivation of vegetables. The lack of awareness among the farmers about the nematode problems and non-availability of suitable package of practices to extension workers for managing the root knot nematodes are the major hindrance for protecting the vegetable crops from root-knot nematodes. Chemical approach of nematode management is no doubt effective but high doses of nematicides required for managing nematodes are neither economical nor environmentally safe.

The infection of root-knot nematode produces characteristic disease symptoms on the below ground root system popularly known as 'root gall' or 'knotted roots'. Different sizes of galls are induced depending on their host and the species of the nematode involved. On cucurbits, the nematode induces large galls, whereas in chilli small size of galls is produced. Usually the infection of M. hapla produces small galls as compared to M. incognita and M. javanica. The size of galls also differs with the level of infection as in case of heavy infection large size or multiple galls or secondary galls develops. Besides galling, forking of taproot in carrot and tubercle on potato tubers are also noticed. Above ground symptoms are non-specific in nature. Infected plants exhibit symptoms of general mineral deficiency, yellowing, stunting, wilting during hotter part of the day, chlorosis, premature shedding of leaves and poor look of plants resulting in low yield. The nematodes are also involved in interaction with other soil borne fungi, bacteria, and viruses and cause serious damage to crops. The interaction of root-knot nematodes is known in many vegetables, fibre, pulses and plantation crops. However, the most common problem is the breakdown of disease resistance and wilting of healthy plants. The most common interaction of root knot nematode with *Ralstonia (=Pseudomonas) solanacearum* is causing "Pseudomonas wilt" in tomato, brinjal and potato.

Nematode Management Options

Root-knot nematodes are polyphagous in nature, having high reproductive potential, and have acquired unique mechanism of survival strategy through laying their eggs in protective gelatinous matrix. Management of root-knot nematode is not easy task under intensive crop cultivation system. Therefore, the idea of keeping the nematode population below the economic damage level by adopting different available tactics is advised to the growers. The young tender seedlings of various crops are very much vulnerable to attack by nematode while the older plants achieve some degree of tolerance. Considering the farmer's suitability, following hygienic cultivation practices of vegetable crops could be suggested for managing root-knot nematodes:

Cultural Practices

Cultural practices are the most effective and economical means of managing insect-pests and disease including nematode problems.

- Two to three summer ploughings (20 cm deep) during the months of May – June at the interval of 15 days expose nematodes, weeds, disease propagules and hibernating stages of insect-pests before the sun and cause their death.
- Inter-cropping with antagonistic plants like marigold (Tagetes spp.) reduces soil population of many soil nematodes including root-knot nematodes. Incorporation of such crop in cropping system either as inter-crop or alternative crop should be considered whenever feasible.
- Crop rotation with resistant varieties or non-host crops like mustard, sesame, maize, wheat etc. are useful to bring down the soil population of nematodes below the damage threshold level.
- Application of organic manure, Farm Yard Manures (FYM) at 18 to 20 t/ha reduces nematode population through their action of released toxic substances, enhanced crop tolerance and encouraging soil microbial antagonists

Resistant Varieties

Plant resistance plays an important role in integrated management of root-knot nematodes diseases, however, availability of resistant varieties of vegetable crops are very few in number and many of them are not acceptable to the farmers for their suitability. Some of the resistance varieties exhibited resistance or tolerance reaction to root-knot nematodes are as given below:

Tomato: SL-120, Hisar Lalit, PNR-7, Hisar N-1, Hisar N-2, Hisar N-3, NT-3, NT-8 NT-12, Ronita, Patriot, PAU-5, Mangla and Karnataka Hybrids.

Chilli: Pusa Jawala, CAP-63, CA-2057, Sindhuri, NP-46 A, Mohini, SP-26, P-6-3, K-235.

Brinjal: Giant of Banaras, Black beauty, Gola, Gachha Baigan, Pbr-91-2, IC-95-13, HOE-101, Red Wonder.

Cowpea: Barasati mutant, 82-IB, C-152, IHR-29-5, GAU-1.

Pea: B-58, C-50.

Potato: Kufri Dewa.

Okra: Kanki local green, Harichickni,Vaishali Badher.

Pumpkin: Dasna, Jaipuri.

Water melon: Shahjanpuri.

Ridge gourd: Panipati, Meerut special.

Chemical Control

Chemical control with the application of nematicides is the most effective means of nematode management. However, most of the effective nematicides have been withdrawn from the world market. At present, a few insecticides having nematicidal property are available to the farmers but because of their high doses required to manage nematode, it becomes cost-ineffective and leaves high pesticide residues to the harvested crops. Despite their inherent drawbacks, chemical nematicides could be applied judiciously so that the doses and cost are reduced drastically. The application of nematicides through bare-root dip treatment, seed treatment and nursery bed treatment has been proved to be effective to protect the young seedlings from nematode attack.

Nursery Bed Treatment

In most of the cases, the infection is carried through infested seedlings from nursery bed. The damage caused by root-knot nematode to the root system of tender seedlings is more harmful than to older plants. The application of nematicides to nursery bed helps to raise nematode free seedlings. Moreover, it reduces the dose of nematicides and cost substantially. The soil application of carbofuran (Furadan 3G) at 0.3 gm a.i./m2 is sufficient for producing nematode free seedlings of many transplanted vegetables crops. The treatment of nursery bed with sebuphos (Rugby 20 WP) or carbofuran (Furadan 3G) or benfurocarb (Oncol 50 WP) at 0.3 or 0.6 g a.i./m2 at the time of sowing reduced root-knot nematode and in seed yield of tomato by 25-62%.

Bare-root Dip Treatment

The seedlings of many transplanted vegetables crops can be dipped in systemic nematicides like oxamyl, prophos, and dimethoate at 500 to 1000 ppm for six hours to denematize the roots. These practices will further ensure to protect the root system of tender seedlings from early attack of nematodes. The seedlings of transplanted vegetables like brinjal, tomato, chilli and planting materials of pointed gourd treated with carbosulfan (Marshal 25 EC) at 500 ppm for six hours provided effective control against root-knot nematodes. Nursery bed treatment with carbofuran at 0.3g a.i./m2 coupled with seedlings deep with carbosulfan 25 EC at 500pm before transplanting effectively manage M. incognita and enhance crop yield in vegetables.

Bare root dip treatment of tomato and brinjal seedlings with Zolone (Phosalone 35 EC) or monocrotophos (Monocil 36 SL) or carbosulfan (Marshal 25 ST or DS) at 0.05% reduces root-knot nematode and increase the yield.

Seed Treatment

The practice of seed-soaking and seed-dressing are important prophylactic measures which give adequate initial protection to the young seedlings of tomato, brinjal, okra, chilli, etc. The most commonly used systemic nematicides viz. fenamiphos, isofenphos, carbosulfan etc. are used at 2-3% w/w.

Seed dressing with carbosulfan (Marshal 25 ST) at 3% w/ w is quite effective for managing root knot nematode in okra, bottle gourd, pointed gourd, bitter gourd and jute. Seed soaking with dimethoate, carbosulfan (Marshal 25 EC) can also be adopted for providing better crop with early protection against nematode.

Field Application

Field application of carbofuran 3G at 2Kg a.i./ha in tomato, brinjal and okra reduces nematode population and increases yields.

Biological Control

Despite its several limitations, biological control of root knot nematode is cost effective and eco-friendly method. As a component of integrated nematode management, biological suppression of root-knot nematode is well documented. Several bioagents have been exploited against this but so far only three bioagents viz. *Paecilomyces lilacinus*, *Psuedomonas fluorescens* and *Pasteuria pene*trans have widely been recognized as effective and promising bioagents. Some formulations of *P. lilacinus* (Bionematon, Yorker) and *Pasteuria penetrans* (Pasutsuria 50 WP), *Trichoderma viride* and *T. harzianum* (Trichostar, Tricho guard, Bioderma, Ecoderma etc) are available in the market for checking root knot nematodes.

Integrated Approach for Root-knot Nematode Management

Individual method of nematode control has either proved ineffective or uneconomical approach against root knot nematodes. Therefore integration of various suitable tactics may be an ecofriendly, economically viable and practically feasible approach for managing nematode problems in crops. The adoption of deep summer ploughing during summer period at fortnightly interval along with organic matter application followed by planting with nematode free seedlings is a feasible approach to reduce nematode population.

Similarly, farmers with their available resources could follow integration of cultural, biological, chemical methods and resistant varieties in suitable combination for each crop cultivation system. Soil solarization/summer ploughing alone as well as in combination with carbofuran 3 G at 2 Kg a.i./ha has been found effective against nematodes infesting brinjal, chilli and tomato. Seedlings raised in solarized nursery beds treated with carbofuran at 0.3g a.i/m2 integrated with application of neem cake at 5q/ha gives better check against nematodes infesting vegetables and increases yield.

Wheat Seed Gall Nematode (*Anguina Tritici*)

This nematode is one of the most serious pests of wheat in the country. It is the oldest known plant parasitic nematode and the nematode alone causes 'ear cockle' disease in wheat and in association with the bacterium, Clavibactor tritici produces 'yellow ear rot' or 'tundu' disease. The host range of this nematode is very few in number and wheat is considered as the most suitable host. Although the control of this nematode is simple and easy as compared to other plant parasitic nematodes, it is still troublesome in many wheat growing parts of Rajasthan, UP, Bihar and Madhya Pradesh particularly in tribal belts where tons of wheat grains are wasted every year. Wheat galls are the primary source of dissemination for nematode either as seed mixture of cockle. The juveniles of A. tritici remain viable in anhydrobiotic state inside the cockle for several years. After sowing the galls/cockles come in contact with soil moisture and become soft leading to release of large number of second stage juveniles. These juveniles infect the growing point of seedlings as ectoparasite and are carried to inflorescence due to the natural growth of seedling. The nematode enters the floral primordia and become endoparasite and eventually the floral primordial converted into seed galls. The initial visible symptom is enlargement of basal stems near the soil surface at the stage of 20 to 25 day-old seedlings. Generally the infested plant exhibits more number of tillers and grows fast as compared to healthy ones. Twisting, curling, crinkling of leaves and stunted growth are common symptoms in the early stage of plants growth the affected ears are typically swollen, broader with or a few awns on the glumes. The cockled ears contain initially green galls later in each spike-let 1 to 5 galls can be seen.

Yellow Ear Rot or Tundu Disease

The early stage of plant exhibits similar disease symptoms as that of ear cockle disease. The yellow ear rot disease is primarily caused by a bacterium, Clavibactor tritici only in the presence of nematode, A. tritici. Under humid climatic conditions, the characteristic symptoms are appeared with a production of bright yellowish bacterial slime on the leaf surface which can be seen trickling down the ears. During dry weather, these slimes become hard, diseased spikes are generally distorted, stunted and narrower than healthy ones with the grains partially or completely converted by bacterial mass.

Management Options

The nematode is easy to manage as because the gall is the only source of ear-cockle and tundu disease. Both physical and mechanical methods are successful for eradicating this nematode from several developed countries of the world, however, in India this nematode is still becoming problems probably due to poor awareness and failure of national campaign against the dreadful disease of wheat.

Physical Methods

- Hot water treatment of wheat seed-lots at 54 to 56 0 C for 10 to 20 minutes.
- Water floatation of seed galls in 5-10% salt solution for 5 to 15 minutes. Seeds containing nematode gall will float in water surface and can be collected and discarded.

Mechanical Methods

Fanning or winnowing is an effective method to remove the galls from seed lots. Sieving/screening is a common practice and most successful method for eradication of seed gall nematode from many countries, though complete removal of gall is not possible with this methods because some large sized galls are retained on the sieves.

Cereal Cyst Nematode (*Heterodera Avenae*)

Heterodera avenae is the causal organism of a serious disease popularly known as 'molya' disease of wheat and barley. This disease was first time recorded from Rajasthan in India and subsequently it is known to occur in major wheat growing states viz. Punjab, Haryana, Uttar Pradesh, Delhi, Himachal Pradesh and Madhya Pradesh of India.The nematode is mainly confined to the family gramminae. The nematode infested fields exhibit patchy growth with stunted and yellowish plant. Infested plants shows thin narrow leaves, reduced tillering, fewer leaves and small size of ear heads with reduced number of grains. The roots of nematode attacked plants appear bushy, bunchy due to emergence of root lets at the site of infection and slight swelling of root tips may be encountered. The above ground symptoms are often confused with the general deficiency symptoms. However, presence of cysts is the only confirmatory evidence for nematode infection.

The second stage juveniles infect the growing tips of roots and upon feeding develop specialized syncytial cells for their growth and development. After three months, the juveniles achieve lemon-shaped sedentary female which is found to attack with roots. Eggs are laid inside the female body and after death of female, the body cuticle transform into brown cyst. During off-season, nematode survives in cysts.

Management Options

Cultural Practices

- Crop rotation with non-hosts like sarson, toria, raya, taramira, gram, berseem, carrot, coriander etc. with wheat.
- Deep summer ploughing (2-3) at an interval of 10-15 days during hot summer months
- Growing wheat cultivar C-306 as trap crop early in October
- Growing Resistant varieties: Barley cultivar like Rajkiran, C-164, BH-72.

Chemical Control

Field application of Carbofuran (FRADAN 3G) at 2kg a.i./ ha have been found effective.

Integrated Management

Integration of different tactics was found economical against cereal cyst nematode. Early sowing in the month of November along with field application of carbofuran at 2kga.i./ha is quite effective in increasing yield and reduction of cyst population in soil.

Potato Cyst Nematode (*Globodera Rostochiensis, G. Pallida*)

It is one of the serious nematode pests of potato in some southern states like Tamil Nadu, Karnataka, and Kerala. The nematode is popularly known as 'golden nematode' and has been recognized as one of the major crop protection problems of the world. The nematodes are responsible for average losses of 9% of global potato amounting to about 40 million tons. In India, potato cyst nematode is known since 1961 when F.G. Jones detected this nematode from a field at Vijayanagaram state farm in Ootacmund of Nilgiri Hills in Tamil Nadu. Considering importance of potato cyst nematode in the country, the Government of Tamil Nadu imposed the Destructive Insect Pest Act 1919(DIP Act 1919) in 1971 to contain the nematode in the Nilgiri Hills.

It is difficult to detect the disease symptoms at low infestation of nematode but the symptoms are prominent only after the build up of population in soil. The disease symptoms appears in small patches of poorly growing plants, temporary wilt of plants in day time, stunted plants, unhealthy yellowish foliage and poor root systems, reduction of number and size of tubers and production of potato yield gradually reduced over the years. The nematode is primarily confined to the family solanaceae and depends on the host root diffusates which induce the hatching of second stage juveniles from eggs. The second stage juveniles infect the root and modify the cells as giant cell for ensuring permanent nourishment to reach adult stage. The adult females are white spherical shape which is found attached with root and after death of female turn into brown cyst. All the eggs laid by the females are retained inside the body. This cyst containing eggs are protected and remain viable for several years in soil even in the absence of potato. The pathotypes RO1 and RO5 of G. rostochiensis and Pa2 and Pa3 of G. pallida of potato cyst nematodes are known to be prevalent in India.

The nematode cyst containing eggs are generally spread through soil particles adhering to tubers, farm implements, gunny bags, farmers' feet etc. However, irrigation water or rain water running down the hill slope carry the cyst from the infested field to uninfested fields.

Management Options

Cultural Practices

- Growing non-host solanaceous vegetable crops like cabbage, cauliflower, beet root, carrots, garlic, radish, turnip etc. One of the most useful rotation as potato-cabbage-carrot is commonly practised by the farmers of Nilgiri Hills.

- Resistant Varieties: Kufri Swarna.

Chemical control: Use of carbofuran (FURDAN 3G) at 2kga.i./ha is effective to reduce nematode population as well as increasing potato yield.

Key Nematode Pests of Rice in India

- Rice root knot nematode, Meloidogyne graminicola
- Rice root nematode, Hirschmanniella spp.
- Rice stem nematode, Ditylenchus angustus
- White tip nematode, Aphelenchoides besseyi
- Rice cyst nematode, Heterodera oryzicola.

Rice Root Knot Nematode (*Meloidogyne Graminicola*)

Rice root knot nematode, Meloidogyne graminicola is a well established nematode pest of rainfed upland rice. It poses serious problems in boro and kahrif nursery particularly in sandy loam or recent alluvial soils of West Bengal. It is also becoming problem in transplanted rice grown in waterlogged conditions. The widespread occurrence of M. graminicola has been found in Assam, West Bengal, Gujarat, Orissa, Karnataka and Tripura.

The above ground symptoms are nonspecific in nature as yellowing, stunting of foliage, delayed flowering by 10 to 15 days and reduced number of tillers. The presence of characteristic 'hook shaped' or 'ring-like' root gall on the root tip of growing rice seedlings is the confirmatory evidence for the association of this nematode. Galls produced by the nematode induce growth of lateral root lets and root hairs. The yield losses due to M. graminicola has been estimated to be between 16 to 32 % in upland rice and in severe cases go up to 64 %. A complete failure of boro rice nursery in 'simurali' in the district of Nadia, West Bengal has been noticed

After harvesting of rice, the nematode may survive in egg stage in soil or continue to reproduce on various weeds. Female often remains concealed within root tissue and eggs are laid in cortical tissues and hatched juveniles reinfect the same roots. It completes life cycle within 19 days at 22 to 29 0C in upland rice.

Management Options

Cultural Practices

- Scheduling crop rotation with non-host crops as cauliflowers, sesame, groundnut, onion, maize, soybean, cowpea
- Weeding: Rice fields supports a diverse kind of weeds like Echinochloa spp., Eleucine indica, Paspalam scrobiculatum, Cyperus spp. etc.
- Resistant varieties: CR-143-2-2, CR-147-2-1, CR-1009, CT-428, Sudha, Murti.

Chemical Approaches

- Seed soaking with carbosulfan (Marshal 25 EC) at500ppm or Carbosulfone at 0.1% for 12hrs
- Bare dipping in carbosulfan (Marshal 25 EC) at 500ppm for 20 minutes
- Nursery bed treatment with carbofuran (Furadan 3G) at 1kga.i./ha and similar dose of nematicides at 7 and 50 days after transplanting.

Rice Root Nematode (*Hirschmanniella* Spp.)

Rice root nematode is a migratory endoparasite of roots and occurs predominantly in moist habitat. Rice root nematode, Hirschmanniella spp. are unique migratory endoparasites of rice and cause yield losses to the extent of 19% in rice in West Bengal. Hirschmanniella oryzae and H. mucronata are two economically species in rice and occurrence of H. gracilis is doubtful in India (per. Comm. Dr. M. R. Siddiqi). The juveniles and adult stages penetrate through entire length of roots and feed on cortical cells leading to the formation of channels or cavities in the roots. Its feeding sometimes extends to central vascular regions. The infected roots exhibit water-soaked brown lesion which are of mostly spindle shaped. The physiological function of infected plants is disrupted and plants growth reduced. The above ground symptoms are non-specific as stunted growth, leaf chlorosis, reduced tillering and delayed flowering.

The population of Hirschmanniella species was found maximum during active growth phages of rice. The population build up of this nematode increases after transplanting up to 80 days of riceand declines when roots of rice starts degenerating.

Rice root nematode survives better in poorly drained clay and heavy soils. It can survive even in high temperature of May-June (35-45 0 C) as well as low temperature of December-January (8-12 0 C) in the North Indian conditions. Their survival in soil is much longer than in roots in flooded soils. H. oryzae can survive more than 12 months in wet soils. A number of weeds found in rice fields serve as alternative host for this nematode. In West Bengal, the nematode can survive in the months of summer in the absence of any crops under laterite soil conditions. Under rice-wheat cropping system, the nematode maintains a very high population, though wheat is not the host for the nematode. Spread of this nematode occurs mainly through irrigation water, flood water, soil adhering to farm implements, field workers and root of rice seedlings.

Management Options

Direct seeding of rice has been found to be more vulnerable to attack by this nematode compared to transplanted crop.

Cultural Practices

- Early planting of rice in the month of June or middle of July
- Use of organic amendments such as mustard cake or neem cake at 220-240 kg/ha
- Balanced NPK fertilization
- Crop rotation with wheat, linseed, potato, cauliflower, mustard and gram in rabi season
- Deep dry summer ploughings
- Weeding during standing rice and in absence of the crop
- Growing resistant varieties/cultivars: TKM-9, CR-142-3-2, CR-52, N-136, W-136
- Sesbania rostrata can be used as trap crop for H. oryzae.

Chemical Approaches

- Nursery bed treatment with carbofuran 3G or phorate 10 G at1.0 kg a.i. /ha followed by I.0 kg/ha at 7 and 50 days after transplanting

- Bare root dip treatment with chlorpyriphos/carbosulfan 25 EC /monocrotophos 36 EC at 1000-2000ppm for 20 to 30 minutes
- Seed soaking with carbosulfan 25 EC or isofenphos at 0.2% for 6 hours.

Rice Stem Nematode (*Ditylenchus Angustus*)

Rice stem nematode is usually problematic in rice grown in deep water situations. The vernacular name 'stem nematode' is derived from the stem inhabiting nature of the nematode. It is an obligate parasite and serious pest of rice causing popular disease symptom referred to as 'Ufra disease'. In Bangladesh and some areas of India, 100 % yield loss has been recorded due to severe attack of D. angustus. The nematode has been found prevalent in Malda, Murshidabad, Hooghly, 24-Parganas, Jalpaiguri, Coochbehar and West Dinajpur districts of West Bengal and Sibsagar, Jorhat, Morrigaon, Sonitpur, Borbeta and Dhubri in Assam

The symptom produced by rice stem nematode is popularly known as Ufra or dakpora disease. The Ufra symptoms appear in patches and subsequently spread to the entire field. The nematode attack at vegetative stage results in yellowing or whitish pattern on the leaf sheath and margin becomes corrugated. In due course of time, the splash pattern turns brownish stains and stem and inter node become black. Twisting of leaf and leaf sheath are commonly found symptoms. Sometimes infested nodes give bushy appearance due to branching. The Ufra symptoms may be grouped as:

Swollen or Thor Ufra: Panicle does not come out, it remains enclosed within the leaf sheath and infected portion tending to branch.

Pucca or Ripe Ufra: Panicle emerges partially and panicle bears filled grains at tip only.

The primary sources of infection D. angustus are rice stubbles, straw, wild rice and weeds found in rice fields. The nematode can overwinter through quiescent state (fourth stage juvenile) which remains viable up to 15 months. They live in coiled anhydrobiotic state in grains; dried plants parts left in the fields and reinfect the crop in the next season.

Cultural Approaches

- Destruction of rice stubbles, weeds and wild rice in the rice fields
- Crop rotation with jute, sesame, mustard with rice
- Summer ploughing helps nematode to destroy by desiccation in the scorching heat of sun
- Growing early (Padmapani) or resistant (Rayada selection) varieties

Chemical Approaches

- Spraying with diazinon at 0.01%
- Soil application of carbofuran 3G or phorate 10G at 1 kg a.i. /ha at transplanting

White Tip Nematodes (*Aphelenchoides Besseyi*)

besseyi is a specialized parasite attacking aerial parts of its natural host, rice. Though rice is the most suitable host of this nematode, it can infect tuberose, onion, soybean, sugarcane, oat, millets, orchids etc. The most characteristic disease symptom is 'white tip' in rice leaf produced by this nematode due to which the common name of the nematode is 'white tip nematode'. It has been recorded in serious form in rice from Gujarat, Tamil Nadu, Madhya Pradesh, Andhra Pradesh and West Bengal. In West Bengal, the nematode is becoming problem in the rice fields adjacent to tuberose fields in different tuberose growing areas of state. The nematode is prevalent through out the state of West Bengal.

It is easy to detect the presence of nematode within the rice seeds. In field, the initial appearance of symptoms as the leaf tip up to 5cm becomes pale yellow or whitish at tillering stage and subsequently leaves get dry. These symptoms are found for a short period in the plant. The tip of the flag leaf are often twisted which may obstruct the emergence of panicles. Infested panicles are shorter and lighter in weight as compared to healthy panicles.

besseyi survives as pre-adults as well as in adult stages (quiescent state) beneath the hull of rice kernel and does not survive in soil after harvesting rice plants. Infected seeds or presence of other alternative hosts help nematode to survive up to next crop. They usually remain in coiled anhydrobiotic state in rice between lemma and palea up to 3 years. The infected rice seed is the only means for rapid spread of *A. besseyi*. It also spread through irrigation water or flood water. Female lays eggs on rice plants. All the developmental stages occur on rice plant. The life cycle of *A. besseyi* is completed within two weeks and, therefore, several generations are completed within a cropping season.

Management Options

- Healthy rice seed checks the spread of white tip disease of rice
- Rice stubbles to be burnt or destroyed after harvesting
- Simple spreading the rice seed on the concrete floor on bright sunny days at least 4 hours for 6 consecutive days kills nematodes inside the grain
- Seed-soaking in water(1:2 ratio) for overnight followed by adding two volume of boiling water for 10 minutes and then drying of seeds in shade
- Seed treatment with carbosulfan 25EC at 0.1% for 12 hours.

Rice Cyst Nematode (*Heterodera Oryzicola*)

Heterodera oryzicola has been reported as a serious pest of rice in Kerala. It has also been known to occur in Burdwan and Bankura district of West Bengal. This nematode is one of the serious pests of rice and banana in Kerala. Its occurrence has been recorded from Karnataka and Goa.

The browning of roots and chlorosis of leaves, retardation in growth, early flowering of plants by 10 to 15 days and partial filling of grains are the typical symptoms of infection of H. oryzicola. No gall developed on rice roots. The presence of brown cyst on rice root is the confirmatory evidence of this nematode infestation. One life cycle is completed in 30 days and 12 generations may occur in a year. Yield losses could be to the extent of 38% due to the attack of nematode. The nematode spread through infested seedlings, irrigation water or farm implements.

Management Options

- Soaking of seeds with phenamiphos at 0.02% for 6 hours
- Soil application of carbofuran or phorate 1kg a.i. /ha at 7 and 50 DAT
- Growing resistant varieties like Lalnakanda, CR143-2-2
- Regulatory measures on the movement of banana rhizomes as well as rice seedlings from infested areas have to be adopted.

Burring Nematode (*Radopholus Similis*)

This nematode is internationally quarantined pest and is capable of parasitizing many fruits, spices and plantation crops. It is known to cause a serious disease of black pepper popularly referred to as 'pepper yellows' in Indonesia and 'slow wilt' in India and 'spreading decline' in Florida. Many economically important crops like banana, citrus, betelvine, coconut, arecanut black pepper, ginger etc. are seriously affected by the nematode. The nematode is mainly problem in Southern states like Kerala, Karnataka, Andhra Pradesh. Recently, it has got spread to the state like Orissa, Manipur, Maharastra, Madhya Pradesh and Himachal Pradesh and Gujarat possibly through indiscriminate movement of planting materials of banana, ginger and turmeric rhizomes.

Nematode infested banana plants exhibits 'toppling disease' at bearing stage, premature defoliation, poor plant vigour and finally reduction of bunch size and weight. In root, lesion develops on root and subsequent rotting and decaying of tender roots due to the involvement of other soil microorganisms. The coconut plants attacked by R. similis show general decline symptoms like yellowing, stunting and smalling of leaves and button shedding resulted in low yield. In black pepper, slow growth of vines, yellowing of leaves, rapid panicle declines followed by severe die-back and death of vines are common symptoms.

The roots of coffee, arecanut, betelvine are seriously attacked and develop characteristic lesions and root decay and rotting. The nematode is a migratory endoparasite of root and feed on the succulent tissues of feeder roots. Due to their intracellular movement, the nematode destroys the cells and forms burrows or cavity inside the root. All the developmental stages are capable of feeding roots. Eggs are laid in root tissues and newly hatched juveniles start feeding and develop inside roots. Therefore, the entire life cycle may be completed within the roots. The total life cycle is completed in 20-25 days. In India, only the banana race of R. similis is prevalent.

Management Options

The nematode is difficult to manage due to its endoparasitic nature and wide host range.

Preventive Measures

- Planting material/seedlings should be free from nematode
- Any occurrence of discolouration on the rhizome should be removed and treated with Bordeaux mixture or nematicides
- Hot water treatment of rhizomes at 50-55 0C for 20-25 minutes could be done to denematize the planting materials.
- Raising of seedlings of coconut, arecanut, black pepper from nematode free nursery bed
- Before planting, sun drying of banana rhizomes is also effective to reduce nematode population

Curative Measures

- Application of carbofuran 6 g a.i. or phorate 3 g a.i./plant 2-3 times in a year is effective to reduce the nematode in coconut and banana.
- Application of neem cake at 400 g/plant once at planting and second dose after 4 months increase bunch weight and the yield of banana
- Avoiding susceptible crops as intercrop
- Intercropping with Crotalaria juncea reduces R. similis population
- Bioagents like *P. lilacinus, Glomus fasciculatum, Pasteuria penetrans* are promising agents against *R. similis.*

Resistant/Tolerant Varieties:

Banana: Kadali, Pedalimoongil, Kunnan, Pey Kunnan, Pisang Seriby

Arecanut: Resistant: VTL-11x VTL-17

Tolerant: Indonesia-6 (VTL-11)

Mahuva-B, Andaman-5 (VTL-29e).

Coconut: Kenthali, Klappawangi, Hybrid Java Giant (JG X Kulasekheram Dwarf Yellow (KDY), KDY X JG, Java Tall X Malayan Yellow dwarf, San ramon X Gangabondan.

Citrus Nematode (*Tylenchulus Semipenetrans*):

Citrus nematode is found in all the citrus growing areas of the country and is widely recognized as economically important pest of citrus. It is one of the causal factors for 'slow decline' in citrus' which is characterized by general reduction of the tree growth, lack of vigour, yellowing of foliage and small size of fruits. The nematode is semi-endoparasite of citrus root.

It causes symptoms that are often non-descriptive and difficult to diagnose. The nematode is often unnoticed in the seedlings in nursery which causes widespread distribution.

The presence of nematode is best confirmed by microscopic observations of soil and root samples. The female nematodes and their gelatinous matrix containing eggs adhering soil particles and give the roots a dirty appearance which is not easily washed off. The most serious effects of the nematode on the growth and yield of citrus are usually encountered when new seedling are planted on old orchard. This condition is known to as 'citrus replant problem'. The young tree grows slowly and fruiting is delayed. This condition of infested trees is referred to as 'slow decline' which implies general deterioration of citrus trees beginning with production of smaller and fewer fruits. The extent of decline in mature trees is related to their vigour, tolerance to nematode and to the degree of infection.

T. semipenetrans feeds on surface layers of roots causing discolouration and necrosis. A young female penetrate deeper root tissues and establishes a feeding site around the head. The feeding site comprising of 16 cortical cells referred to as 'nurse cell'. The posterior part of mature female body remains outside and eggs are laid in a gelatinous matrix outside the host tissue. The life cycle of this nematode is completed within 6-8 weeks under optimum temperature at 25 to 31 0C.

Prophylactic Measures

- Nursery raising of seedling should be free from nematode infection
- Previously infested citrus orchard should be either avoided or fumigated to kill any nematode population in soil.
- Preventing run off water from adjacent infested citrus orchard
- Use of clean equipment/implements for cultural practices because movement of any adhering soil particles from one place to another in orchard may disseminate the nematode.

It is difficult to eliminate the nematode once it is established in orchard. Therefore, regular monitoring is essential for preventing the nematode to reach above the damage threshold level. Usually the nematode at low population level (500/g of feeding root) are not damaging to the crop but at high population (4000/g root) causes devastating damage to citrus plant.

Curative Measures

- Application of oilcakes of neem, mahua, groundnut etc. at 1kg/plant can reduce nematode populations.
- Combined use of neem cake at 1kg/plant along with carbofuran 3G (Furadan) 1.0 kg a.i./ha is also effective.
- Use of biocontrol agent like Paecilomyces lilacinus at 4 g/plant along with carbofuran 3G (Furadan) 1 kg a.i./ ha gives good results.
- Use of resistant 'trifoliate' citrus stock may be an efficient method to check the nematode.

Cultural Practices: Removal of old feeder roots before the start of growth flush followed by application of FYM helps to reduce nematode population in soil.

Foliar Nematode (*Aphelenchoides Besseyi*) of Tuberose

Tuberose is commercially important for loose flower, cut flower and for extraction of essential oils. Its commercial cultivation in West Bengal was started in the second half of 19th century to meet aesthetic need of the English people harboured in Calcutta. At present the total area under tuberose in India is estimated to be about 20,000 ha, out of which 2,110ha is in West Bengal and it is mainly confined in Kolaghat-Panskura, Ranaghat and Haringhata areas. In terms of area and production of tuberose, West Bengal ranked first. Recently, foliar nematode, Aphelenchoides besseyi has been appeared as a serious problem in tuberose in West Bengal. This nematode was first time reported from Hawaii Islands on the leaves of tuberose. The occurrence of floral malady caused by A. besseyi in tuberose was recorded from Ranaghat areas of Nadia district of West Bengal. In India, this nematode is widespread in Eastern and Southern states particularly in rice causing 'white tip disease' and estimated to cause yield loss 20 to 30% in rice, though floral disease caused by A. besseyi in tuberose was not encountered any other states except West Bengal. Recently, it has got spread to neighbouring state, Orissa either through movement of bulbs or other means. Recent survey results revealed that A. besseyi is a major limiting factor for cultivation of tuberose in Ranaghat and Haringhata regions of Nadia, Bangaon of 24-Parganas (South) and some pockets of Howrah and Midnapore districts of West Bengal. The high population of *A. besseyi* was also recorded from fields of Kolaghat-Panskura-II of Midnapore district. The 'single' cultivar of tuberose was found to be the most vulnerable to damage caused by *A. besseyi* as compared to 'double' cultivar. Khan et al. (1999, 2001) investigated for severe foliar disease infestation in tuberose and found A. besseyi is the primary causal agent for malformed flowers. The population of A. besseyi causing white tip disease in rice is the same population infecting tuberose and causing floral disease. *A. besseyi* is now recognized as the key nematode pest of tuberose and it was becoming a potential threat for cultivation of tuberose in West Bengal.

The infected flower stalk initially appears rough, stalk become crinkled, stunted and finally distorted and in severe cases flower buds failed to bloom. Brown streaks appear on leaf bracts and petals and subsequently develop rusty brown spots. The severely infected flower stalk becomes rotten and brittle over drying, even get blind. The number of flowers per stalk is also reduced and small crinkled and distorted flowers which are not acceptable in the market. The nematode, *A. besseyi* remains in masses forming 'nematode wool' which could be easily recovered from dark brown spots. The ovary contains large number of nematodes. This nematode is more serious during rainy season generally from July to September and cent percent loss of second year crop particularly in 'Calcutta single' cultivar of tuberose was recorded. However, in 'Calcutta double' cultivar 30% to 40% flower stalk rendered unsaleable and from individual flower stalk up to 45,000 nematodes were recovered. In Calcutta single, the nematode could cause 38 to 59% yield loss

The dissemination of *A. be*sseyi occurs through bulbs collected from infested fields. The nematode survives in coiled anhydrobiotic condition (quiescent pre-adult and adult stages) in the scaly leaves outside the bulbs. The nematode can also survive in the dried scaly leaves, stems and flowers more than 25 months; however, they can not survive in soil for long time.

Reniform Nematode (*Rotylenchulus Reni*formis)

The adult female of reniform nematode (Rotylenchulus reniformis) is an obligate, sedentary semi-endoparasite of a wide range of food, fibre, oilseed, oilseed, fruits and plantation crops. The common name 'reniform' was derived from the kidney-shaped mature female. It has worldwide distribution and is receiving importance as national pest of crops. At present, there are 10 species of reniform nematode known worldwide but R. reniformis is the most widespread and has economic importance. In West Bengal, R. reniformis has been known to be associated with many vegetable crops, banana, tuberose, tea, pulses, fruits, betelvine etc.

The symptoms of damage to crops are non-specific on the above ground or even in the below ground parts necessitating a close observation to confirm their presence and damage. It feeds on cortical tissue, phloem and pericycles and its infection may cause formation of necrosis on roots of certain crops. Symptoms appear as root discolouration, shedding of the leaves and formation of malformed fruits and seeds. In addition to causing direct damage to plants roots, the nematode in concert with other pathogens like Fusarium spp., Verticillium spp., Sclerotium rolfsii and Rhizoctonia solani develop diseases complexes. It has also been reported to parasitize the bacterial nodules.

The nematode is capable of surviving in air dried soil for a long period of time. The retention of moulted cuticles of previous stages is a unique adaptation for survival of nematode in soil. Individual young females, males and fourth stages juveniles could survive in a coiled anhydrobiotic state with encrusted cuticles in soil. Survival of this nematode inversely related to the rate of moisture loss in soil. Therefore, alternate drying and wetting of soil resulted in sharpe decline of population density of the nematode in soil.

The first moult occurs within eggs and eggs are hatched in water without the influence of root exudates. Juveniles develop to pre-adult stage without any feeding host tissue and quickly completing three superimposed moulting. The young female is the only infective stage. After infection to the roots, young female orient themselves perpendicularly to the longitudinal axis of roots with the posterior part remain outside the root. After establishing the feeding site, it develops into kidney shaped female with posterior portion protrude outside the root. Egg laying starts within 7-10 days after invasion and eggs are laid into a gelatinous matrix secreted by six specialized cells around vagina. Each egg mass contains 30-200 eggs. Total life cycle is completed within 3 to 4 weeks depending upon temperature and host suitability.

Cultural Methods

Crop rotation with non hosts crops like mustard, maize, sugarcane, marigold. Growing the susceptible crops in winter seasons in the multiple cropping systems

Organic Managements

Application of FYM, oilcakes like neem, karanj, mustard etc. have been found promising.

Summer ploughing: Two to three summer ploughings during hot months

Irrigation management: Irrigation between ploughings results in alternate drying and wetting which may stimulate exsheathment of young female of R. reniformis

Crop Husbandry: Good crop cultivation practices like field preparation, fertilizer application and moisture management can improve crop tolerance to nematodes

Resistant Varieties:

Cowpea: Pusa phalguni, C-152, RC-48

Papaya: Solo, Washington, Coorg Honey Dew

Onion: Evergreen

Chickpea: BG-425, BG-426, BG-434, BG-268, BG-273

Blackgram: UG-201, UG-135

Chilli: Pusa Jawala

Chemical Control: Soil application of carbofuran (Furadan 3G) at 2 kg a. i./ha gives good control but it may not often be economical in many low value crops. However, judicious use of nematicides may be adopted by restricting their use at nursery bed, seed treatments, bare dipping of vine cuttings and pit application particularly for transplanted crops.

Lesion Nematodes (*Pratylenchus* Spp.)

Pratylenchus spp. are migratory endoparasite of root. The vernacular name of 'root lesion nematode' is derived from the discoloured patches (lesion) develop on roots.

It has a very wide host range including important crops like wheat, maize, cotton, potato, rice, banana, tea, vegetables, ornamentals and fruits. Some nematode species like Pratylenchus thornei in wheat, soybean, chickpea, sunflower and opium, *P. zeae* in maize, *P. ind*icus in rice, *P. loosi* in tea, *P. coffeae* in coffee and banana, *P. pratensis* and *P.vulnus* in fruits are serious problems.

The above ground symptoms caused by the nematode are non-specific in nature. The nematode usually infects in roots, rhizomes or tubers. Having penetrated into roots, they multiply in large numbers. All the stages of this nematode are infective.

The attacked plant's root exhibits dark red brown lesions caused by necrosis of the invaded cells. Root lesion is the most characteristic symptoms. The lesions initially appear as small elongate, water-soaked spots which soon turn brown to black. Loss of primary roots, pruning or decay of roots, reduced size of blossoms, shrinking of grains are also associated with the nematodes. Several secondary soil borne fungi, bacteria are also involved for rotting and decay of roots and thus normal functioning of infested roots are heavily impaired.

Management options

1. Summer ploughing of field reduces nematode populations
2. Application of carbofuran (Furadan 3G) at 1 kg a.i./ha at sowing reduces crop damage caused by soil nematodes
3. Growing antagonistic crop like marigold (Tagetes patula) cv. Harmony in autumn after main crop or in between rows of main crop.
4. Hot water treatment of bulbs, corms, tubers and fleshy roots can kill the dormant nematodes inside the root.

Pigeon Pea Cyst Nematode (*Heterodera Cajani*)

Heterodera cajani is the only species of cyst nematodes parasitizes a large number of leguminous crops. This nematode is prevalent and gaining importance in almost all pigeon pea growing states. Some populations of this nematode are known to attack sesame also.

The species is distinctive for having large egg-sac (almost double of its cyst size). The nematode completes its life cycle in 16 days at 29°C and in 45-80 day at 10°C to 24°C. It can also reproduce parthenogenetically, though it is bisexual species.

During a cropping season, it can quickly multiply and build up a huge population. At seedling stage of plant pearl-like or lemon-shaped white female can be found attached with roots. Infected plants show yellowing, stunting, poor vigour and pod formation. Several crops like pigeonpea, cowpea, mungbean, soybean, blackgram and sesame are seriously damaged by the nematode.

Ectoparasitic Nematodes

Several ectoparasitic nematodes are emerging as new problems of crops. In spite of being numerous group in soil and attained much adaptive biological features of survival and multiplication as compared to endoparasites, they are regularly ignored as nematode pest.

Many genera like *Tylenchorhynchus, Hoplolaimus, Helicotylenchus, Paratylenchus, Hemicriconemoides, Hemicycliophora, Criconemoides, Xiphinema, Longidorus, Trichodorus* etc. are prevalent in the rhizosphere of agricultural, horticultural and forest crops and their pathogenic potential have been proved and their damaging nature are well documented. However, ectoparasitic nematodes are given least attention as because of their symptoms induced on plant are not easily convincing, often confused with other soil problems and soil pathogens. The role played by this group of nematode is much more dangerous particularly when they interact with other soil microorganisms and thereby make the plants vulnerable to weak pathogens. In modern agricultural system, intensive and extensive cultivation of same crops, changes in cultivation practices like high yielding varieties, poor organic nutrition in soil, indiscriminate use of agrochemicals etc. resulted shift in pest status.

For instances, *Tylenchorhynchus brevilineatus*, an polyphagy ectoparasite of diverse crops has appeared as serious problem as '*kalahasti malady*' disease of groundnut in Andhra Pradesh accounting for 20 to 60% yield losses. This nematode is also reported to cause concern in tobacco in Gujarat.

There are many instances like *T. brassicae* in cabbage and cauliflower, *Hoplolaimus indicus* in rice and jute, *Helicotylenchus multicintus* in banana, *Paratylenchus* and *Criconemoides* in apple and peach, Xiphinema basiri and Hemicriconemoides in citrus and grape and Paralongidorus in sal are emerging nematode problems in the changing agricultural scenario in India.

Future Approaches and Conclusion

Plant parasitic nematodes constitute one of the major limiting factors for cultivation of crops. The changes in agricultural situations have tremendous effects on the emergence of new nematode problems in India. The recent outbreak of M. graminicola in the Mandya district of Karnataka, West Bengal, Orissa, Assam, floral malady caused by Aphelenchoides besseyi in tuberose in West Bengal and Orissa, kalahasty malady in groundnut in Andhra Pradesh, Meloidogyne indica in kagzi lime in Gujarat and Pratylenchus zeae in maize are few examples of serious concern.

Despite the fact, plant parasitic nematodes are mostly neglected and considered as low priority factor for crop production and protection In India. The economic significance of nematodes in agriculture is very often under estimated and their damage potential is not recognized by the plant protection specialists, scientists and administrators.

As an important component of integrated pest management, nematode pathogens cannot be ignored and they could be tackled with the intelligent planning of nematode suppressive crop sequences, summer ploughing, organic manuring, clean cultivation, adjusting sowing time, water and irrigation management and sensible use of nematicides. The increasing concern on the ill effects of chemical pesticides on environment has driven the recent research interests on the use of several alternative strategies like botanicals, biopesticides and cropping system research for management of insect-pests diseases including nematodes. Some of the successes have been obtained for managing plant parasitic nematodes with neem based formulations, fungal formulation of P. lilacinus, bacterial formulations of Pasteuria penetrans and Pseudomonas fluorescens.

However, wide application of bioagents in field scale is still limited probably due to their inconsistent efficacy in different agro-ecological situations. Integration of two or more methods can be explored on the basis of their compatibility, economic viability and availability to the farmers. Developing holistic approaches for managing field problems including nematodes inducing disease complexes in concert with other pathogens like fungi, bacteria and viruses will be the major future areas of research thrust.

Chapter 2

Nematode Basics

The "roundworms" or "nematodes" (phylum Nematoda) are the most diverse phylum of pseudocoelomates, and one of the most diverse of all animals. Nematode species are very difficult to distinguish; over 80,000 have been described, of which over 15,000 are parasitic. It has been estimated that the total number of described and undescribed roundworms might be more than 500,000. Unlike cnidarians or flatworms, roundworms have a digestive system that is like a tube at both ends. Nematodes have successfully adapted to nearly every ecological niche from marine to fresh water, from the polar regions to the tropics, as well as the highest to the lowest of elevations. They are ubiquitous in freshwater, marine, and terrestrial environments, where they often outnumber other animals in both individual and species counts, and are found in the locations as diverse as Antarctica and oceanic trenches. They represent, for example, 90% of all life on the seafloor of the Earth. Their many parasitic forms include pathogens in most plants and animals (including humans.) Some nematodes can undergo cryptobiosis.

Taxonomy and Systematics

At the origin, the taxon *Nematoidea* applied to both roundworms and horsehair worms. Along with *Acanthocephala, Trematoda* and *Cestoidea*, it formed the group *Entozoa*. The first differentiation of roundworms from horsehair worms, though erroneous, is due to von Siebold (1843) with orders *Nematoidea* and *Gordiacei* (*Gordiacea*). They were classed along with *Acanthocephala* in the new phylum *Nemathelminthes* (today obsolete) by Gegenbaur (1859). Then the taxon

Nematoidea has been promoted to the rank of phylum by Ray Lankester (1877) including the family *Gordiidae* (horsehair worms). In 1919, Nathan Cobb proposed that roundworms should be recognized alone as a phylum. He argued that they should be called nema(s) in English rather than "nematodes" and defined the taxon Nemates. For ITIS, the taxon *Nematoda* is invalid. Since Cobb was the first to exclude all but nematodes from the group, the valid taxon should be *Nemates* Cobb 1919 or *Nemata* Cobb 1919.

Phylogeny

The relationships of the nematodes and their close relatives among the protostomian Metazoa are unresolved. Traditionally, they were held to be a lineage of their own, but in the 1990s it was proposed that they form a clade together with moulting animals such as arthropods. This group has been named Ecdysozoa. However, the monophyly of the Ecdysozoa was never unequivocally accepted: while most researchers however consider at least the placement of arthropods as more distant relatives of annelids — with which they were formerly united— to be warranted, the presumed close relationships of the nematodes and relatives with the arthropods has been a major point of contention.

Even though the amount of data since accumulated in regard to this problem is staggering, the situation seems if anything less clear these days. DNA sequence data, initially strongly supporting the Ecdysozoa hypothesis, has become rather equivocal on ecdysozoan monophyly, and is simply unable to refute either a close or a more distant relationship between the arthropod and nematode lineages. That the roundworms have a large number of peculiar apomorphies and in many cases a parasitic lifestyle confounds morphological analyses. Genetic analyses of roundworms suggest that — as is also indicated by their unique morphological features — the group has been under intense selective pressure during its early radiation, resulting apparently in accelerated rates of both morphological and molecular evolution. Furthermore, no distinctive apomorphies of Ecdysozoa are known; even moulting has recently been confirmed to occur outside the presumed clade.

Conversely, the identity of the closest living relatives of the Nematoda has always been considered to be housefly due to its internal structure and its sexual reproduction. Morphological characters and molecular phylogenies agree with placement of the roundworms as sister taxon to the parasitic horsehair worms; together they make up the Nematoida. Together with the Scalidophora, the Nematoida form the Introverta. It is entirely unclear whether the Introverta are, in turn, the closest living relatives of the enigmatic Gastrotricha; if so, they are considered a clade Cycloneuralia, but there is much disagreement both between and among the available morphological and molecular data. The Cycloneuralia or the Introverta — depending on the validity of the former — are often ranked as a superphylum.

Nematode Systematics

Due to the lack of knowledge regarding many nematodes, their systematics is contentious. Traditionally, they are divided into two classes, the Adenophorea and the Secernentea, and initial DNA sequence studies suggested the existence of five clades:

- Dorylaimia
- Enoplia
- Spirurina
- Tylenchina
- Rhabditina.

As it seems, the Secernentea are indeed a natural group of closest relatives. But the "Adenophorea" appear to be a paraphyletic assemblage of roundworms simply retaining a good number of ancestral traits. The Enoplia do not seem to be monophyletic either but to contain two distinct lineages. The old group "Chromadoria" seem to be another paraphyletic assemblage, with the Monhysterida representing a very ancient minor group of nematodes. Among the Secernentea, the Diplogasteria may need to be united with the Rhabditia while the Tylenchia might be paraphyletic with the Rhabditia.

The understanding of roundworm systematics and phylogeny as of 2002 is summarised below:

Phylum Nematoda

- Basal order Monhysterida
- Basal subclass/Class Enoplia
- Basal subclass/Class Dorylaimia
- Class Secernentea
    - Subclass Diplogasteria (disputed)
    - Subclass Rhabditia (paraphyletic?)
    - Subclass Spiruria
    - Subclass Tylenchia (disputed)
- "Chromadoria" assemblage.

Reproduction

Reproduction is usually sexual. Males are usually smaller than females (often much smaller) and often have a characteristically bent tail for holding the female for copulation. During copulation, one or more chitinized spicules move out of the cloaca and are inserted into genital pore of the female. Amoeboid sperm crawl along the spicule into the female worm. Nematode sperm is thought to be the only eukaryotic cell without the globular protein G-actin.

Eggs may be embryonated or unembryonated when passed by the female, meaning that their fertilized eggs may not yet be developed. In free-living roundworms, the eggs hatch into larva, which eventually grow into adults; in parasitic roundworms, the life cycle is often much more complicated.

Nematodes as a whole possess a wide range of modes of reproduction. Some nematodes, specifically *Heterorhabditis* spp., undergo a process called *endotokia matricida*: intrauterine birth causing maternal death. Some nematodes, like other animals (for example segmented worms), are hermaphroditic. The hermaphroditic nematode keeps its self-fertilized eggs inside its uterus until they hatch. The juvenile nematodes will then ingest the parent nematode. This process is significantly promoted in environments with a low or reducing food supply. The nematode model species *Caenorhabditis elegans* and *C. briggsae* exhibit androdioecy, which is very rare among animals. The single genus *Meloidogyne* (root-knot nematodes) exhibit a range of reproductive modes including sexuality (amphimixis), facultative sexuality, meiotic parthenogenesis (automixis) and mitotic parthenogenesis (apomixis).

Free-living Species

In free-living species, development usually consists of four molts of the cuticle during growth. Different species feed on materials as varied as algae, fungi, small animals, fecal matter, dead organisms and living tissues. Free-living marine nematodes are important and abundant members of the meiobenthos. They play an important role in the decomposition process, aid in recycling of nutrients in marine environments and are sensitive to changes in the environment caused by pollution. One roundworm of note is *Caenorhabditis elegans*, which lives in the soil and has found much use as a model organism. *C. elegans* has had its entire genome sequenced, as well as the developmental fate of every cell determined, and every neuron mapped.

Parasitic Species

Nematodes commonly parasitic on humans include ascarids (*Ascaris*), filarids, hookworms, pinworms (*Enterobius*) and whipworms (*Trichuris trichiura*). The species *Trichinella spiralis*, commonly known as the trichina worm, occurs in rats, pigs, and humans, and is responsible for the disease trichinosis. *Baylisascaris* usually infests wild animals but can be deadly to humans as well. *Haemonchus contortus* is one of the most abundant infectious agents in sheep around the world, causing great economic damage to sheep farms. In contrast, entomopathogenic nematodes parasitize insects and are considered by humans to be beneficial.

One form of nematode is entirely dependent upon fig wasps, which are the sole source of fig fertilization. They prey upon the wasps, riding them from the ripe fig of the wasp's birth to the fig flower of its death, where they kill the wasp, and their offspring await the birth of the next generation of wasps as the fig ripens.

Plant parasitic nematodes include several groups causing severe crop losses. The most common genera are *Aphelenchoides* (foliar nematodes), *Ditylenchus*, *Globodera* (potato cyst nematodes), *Heterodera* (soybean cyst nematodes), *Longidorus*,

Meloidogyne (root-knot nematodes), *Nacobbus*, *Pratylenchus* (lesion nematodes), *Trichodorus* and *Xiphinema* (dagger nematodes). Several phytoparasitic nematode species cause histological damages to roots, including the formation of visible galls, which are useful characters for their diagnostic in the field. Some nematode species transmit plant viruses through their feeding activity on roots. One of them is *Xiphinema index*, vector of GFLV (Grapevine Fanleaf Virus), an important disease of grapes. Other nematodes attack bark and forest trees. The most important representative of this group is *Bursaphelenchus xylophilus*, the pine wood nematode, present in Asia and America and recently discovered in Europe.

Nematodes in Agriculture

Depending on the species, a nematode may be beneficial or detrimental to plant health. From an agricultural perspective, there are two categories of nematode: predatory ones, which will kill garden pests like cutworms, and pest nematodes, like the root-knot nematode, which attack plants. Predatory nematodes can be bred by soaking a specific recipe of leaves and other detritus in water, in a dark, cool place, and can even be purchased as an organic form of pest control.

Rotations of plants with nematode resistant species or varieties is one means of managing parasitic nematode infestations. For example, marigolds, grown over one or more seasons (the effective is cumulative), can be used to control nematodes. Another is treatment with natural antagonists such as the fungus *gliocladium roseum*. Chitosan, a natural biocontrol the elicits plant defence responses to destroy parasitic cyst nematodes on roots of sobyean, corn, sugar beets, potatoes and tomatoes without harming beneficial nematodes in the soil.. Furthermore soil steaming is an efficient method to kill nematodes before planting crop. CSIRO has found that there was 13-to 14-fold reduction of nematode population densities in plots having Indian mustard green manure or seed meal in the soil. Hundreds of *Caenorhabditis elegans* were featured in a research project on NASA's STS-107 space mission (which ended in the Space Shuttle Columbia Disaster).

Plant nematodes attack all crops grown in Florida, causing farmers millions of dollars in crop loss annually. Nematodes also are of great concern to the home owner, since they cause severe damage to turfgrasses, ornamentals and home gardens. We are often unaware of losses caused by nematodes, since they are hidden from sight and much of the damage caused by them goes unreported or is attributed to other causes.

Morphology and Anatomy

Nematodes that attack plants are worms, mostly microscopic in size, ranging from 0.25 mm to 3.0 mm (1/100-1/8 inch) long. They are generally cylindrical in shape, tapering toward the head and tail. Females of a few species lose their worm shape as they mature, becoming greatly enlarged in diameter and assuming varying forms, such as pear, lemon, or kidney shapes.

In spite of their small size, nematodes are complex in organization. Plant parasitic nematodes possess all of the major organ systems of higher animals except respiratory and circulatory systems. The body is covered by a multi-layered cuticle which bears surface marks which are used when identifying nematode species.

The cuticle of most is transparent, so that sufficient anatomical detail can be seen with the aid of a low-power "dissecting" binocular microscope, generally 20 to 60 X, to identify most nematode specimens to genus. Much higher magnification (900 X or more) is often needed to identify species.

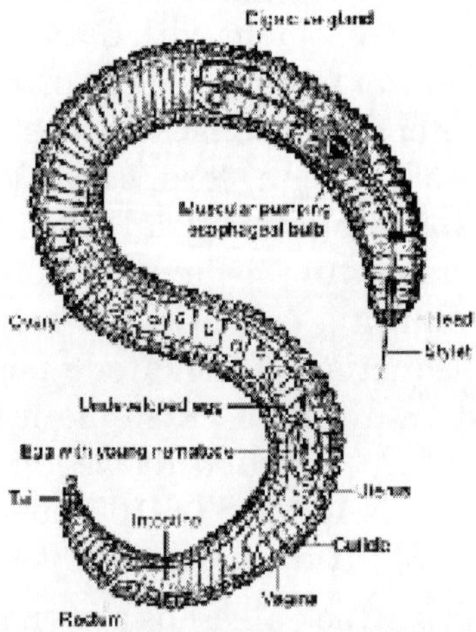

Figure: *Diagram of a typical plant-parasitic nematode.*

Life Cycle and Reproduction

Plant parasitic nematodes have a simple life cycle of six stages: egg, four juvenile stages, and adult. The embryo develops inside the egg to become the first-stage juvenile. The first-stage juvenile molts inside the eggshell to become a second-stage juvenile, which hatches from the egg, and in most species must feed before continuing to develop. The nematode molts three more times to become a fully developed adult.

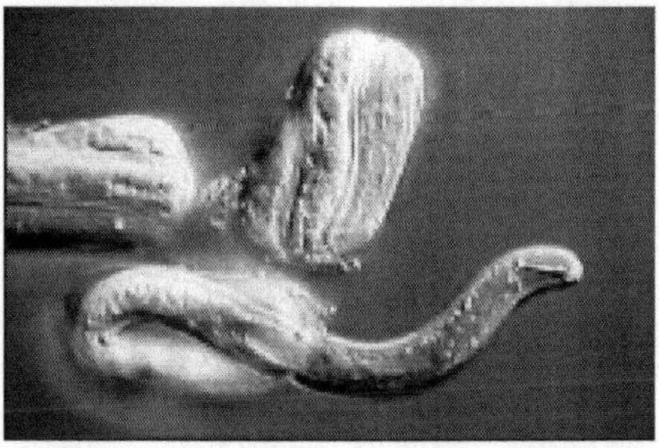

Figure: *Second-stage juvenile nematode hatching from its egg.*

Male and female nematodes occur in most species, and both may be required for reproduction. However, reproduction without males is common, and some species are hermaphroditic ("females" produce both sperm and eggs). Egg production by the individual completes the cycle. The number of eggs deposited by a female varies among species and is affected by their habitat. Most species produce between 50 and 500 eggs, but a few occasionally produce several thousand eggs per female.

The length of the life cycle varies considerably depending on nematode species, host plant, and the temperature of the habitat. During summer months when soil temperatures are in the 80's, many plant nematodes complete their life cycles in about 30 days.

Nematode Feeding and Host-Parasite Relationships

Plant parasitic nematodes feed on living plant tissues. All have some form of oral stylet or spear, which is used somewhat like a hypodermic needle to puncture the host cell wall. Many (probably all) plant nematodes inject enzymes into the host cell before feeding. These enzymes partially digest the cell contents before they are sucked into the gut. Most of the injury that nematodes cause to plants is related in some way to the feeding process.

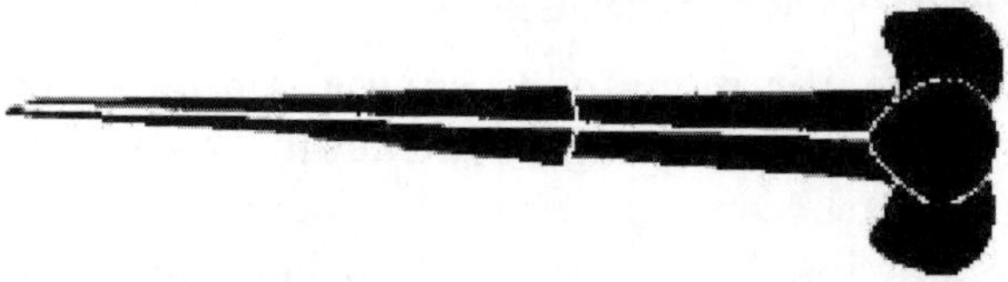

Figure: *A plant-parasitic nematode stylet is used to puncture plant cells, to inject enzymes, and to ingest cell contents.*

Nematodes may feed on plant tissues from outside the plant (ectoparasitic) or inside the tissues (endoparasitic). If the adult female moves freely through the soil or plant tissues, the species is said to be "migratory." Species in which the adult females become swollen and permanently immobile in one place in or on a root are termed "sedentary".

The feeding/living relationships that nematodes have with their hosts affect sampling methods and the success of management practices. Ectoparasitic nematodes which never enter roots may be recovered only from soil samples. Endoparasitic nematodes often are detected most easily in samples of the tissues in which they feed and live (burrowing and lesion nematodes), but some occur more commonly as migratory stages in the soil (root-knot and reniform nematodes).

Those stages of endoparasites which are inside root tissues may be protected from nematicides which do not penetrate into roots, such as some soil fumigants. Root tissues may also shield them from many microorganisms which attack nematodes in the soil. Ectoparasites are fully exposed to pesticides and natural control agents in the soil.

Ectoparasitic nematodes are generally migratory. Most feed superficially at or very near the root tip or on root-hairs, but a few have stylets long enough to enable them to feed deeper in the root. Those which cause the most widespread and severe plant injury in Florida are the sting (*Belonolaimus* spp.), stubby root (*Trichodorus* spp.), and awl (*Dolichodorus* spp.) nematodes. These feed at or near root tips and usually inhibit root elongation.

Ectoparasites which rarely cause severe injury to their plant hosts include ring (*Mesocriconema* spp.) and spiral (*Helicotylenchus* spp.) nematodes. They apparently feed primarily on root-hairs and superficial cortical tissues and cause serious injury only to plants that are especially sensitive to drought stress.

Figure: *A ring nematode (thicker nematode) and spiral nematodes (thinner nematodes) feeding on a plant root. These nematodes are ectoparasitic nematodes.*

Among plant nematodes, only stubby-root nematodes and their close relatives, the dagger (*Xiphinema* spp.) and needle (*Longidorus* spp.) nematodes, are known to transmit plant viruses. The corky ringspot disease of potatoes, a problem in the Hastings area, is caused by a virus which is carried and transmitted by stubby-root nematodes.

Migratory Endoparasites

Migratory endoparasites can move into, through, and out from host tissues at any stage of development (except the egg). Migratory endoparasites generally live and feed in tender tissues such as the root cortex. They burrow through the tissue, breaking open many cells after feeding on them. Cells surrounding the feeding area are often killed by toxic materials from the disrupted cells. The relatively large areas of dead cells usually turn brown, to become small spots or lesions big enough to see, and are often easily colonized by fungi. Root rot diseases are often associated with infestations of migratory endoparasitic nematodes.

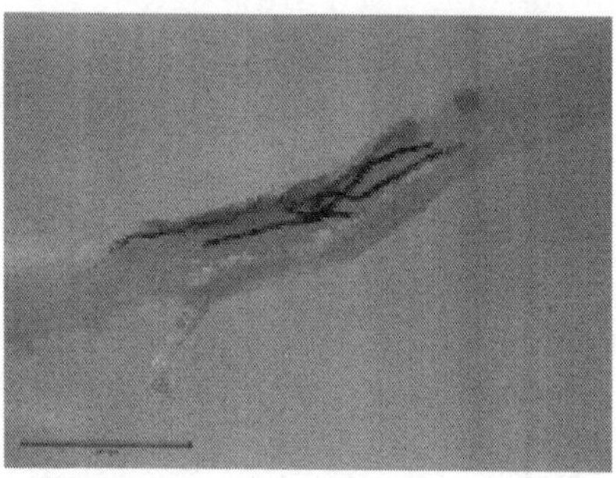

Figure: *Endoparasitic nematodes burrowing within a root.*

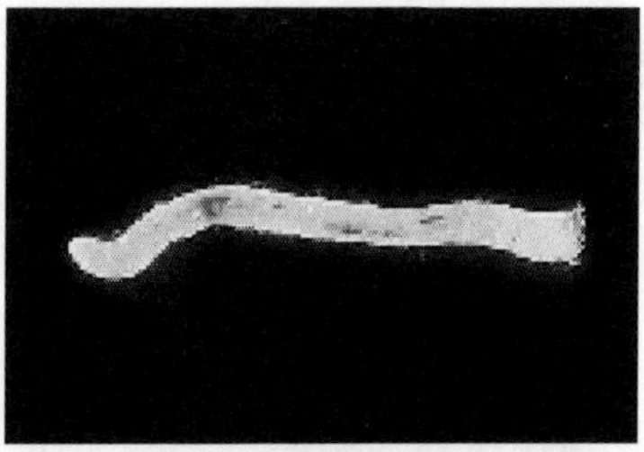

Figure: *Root lesions caused by migratory endoparasitic nematodes.*

Figure: *Rotting palm roots resulting from infection with migratory endoparasitic nematodes.*

The most important examples in Florida are species of burrowing and lesion nematodes. The citrus burrowing nematode, *Radopholus similus*, causes the spreading decline disease of citrus. It is the subject of strict (and expensive) quarantine regulations for ornamentals, nursery stock, and other growing plants being exported from the state, and can severely limit growth of many ornamental plants. Various species of lesion nematodes live in the roots of most crops grown in the state and may be especially damaging to citrus, peanuts, and commercial fern nurseries.

The foliar (*Aphelenchoides* spp.) nematodes are the most important representatives in Florida of a group of migratory nematodes which attack plants at the soil line or above the ground. They feed on or inside the leaves and buds of ferns, strawberries, chrysanthemums, and many kinds of foliage ornamentals, and cause distortion or death of buds, leaf distortion, or yellow to dark brown lesions between major veins of leaves. Other nematodes which attack plants above ground cause leaf or seed galls and still others cause deterioration of the bulbs and necks of onions and their relatives, but are not common here.

Figure: *Symptoms of foliar nematode damage are often angular shaped leaf spots.*

Ectoparasitic and migratory endoparasitic nematodes generally deposit their eggs singly as they are produced, wherever the female happens to be in the soil or plant.

Sedentary Endoparasitic Nematodes

Sedentary endoparasitic nematodes are typified by the root-knot (*Meloidogyne* spp.), cyst (*Heterodera* spp.), reniform (*Rotylenchulus* spp.), and citrus (*Tylenchulus semipenetrans*) nematodes. In most of these species, the second-stage juvenile is the "infective" stage, which moves through the soil. Second-stage juveniles locate host roots and enter them.

They then establish a suitable feeding site within the root tissues. Once a feeding site is selected, the nematode injects growth regulating substances into the cells near its head, causing some of those cells to enlarge. These "giant" or "nurse" cells become specialized food sources for the nematode. At the same time, the nematode becomes immobile, and the body swells to a round, lemon, kidney, or ovoid form.

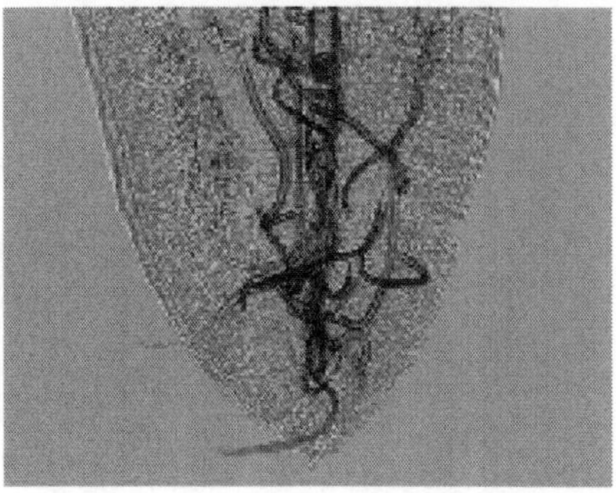

Figure: *Second-stage juvenile nematodes entering a root tip.*

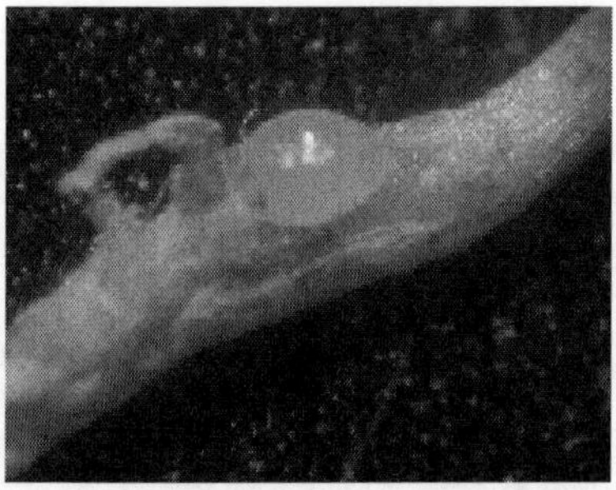

Figure: *Root tissue pulled-back to reveal a swollen root-knot nematode within.*

Mature females of the sedentary endoparasitic nematodes generally produce large numbers of eggs which remain in their bodies or accumulate in masses attached to their bodies. The nematodes and the giant cells on which they feed are very dependent on each other: if the nematode dies, the giant cells die or lose their highly active condition; if the giant cells die, the nematode dies of starvation, because it cannot move to a new site. Sedentary endoparasites damage their hosts by redirecting use of large amounts of energy and nutrients from normal activities to development of the nematodes and their special feeding sites.

The altered tissues of the feeding site also disrupt the vascular system. Roots severely galled by root-knot nematodes also usually deteriorate much earlier from root rots than roots which are not galled: gall tissues are succulent, poorly protected from invasion, and rich in nutrients to help the fungi grow rapidly.

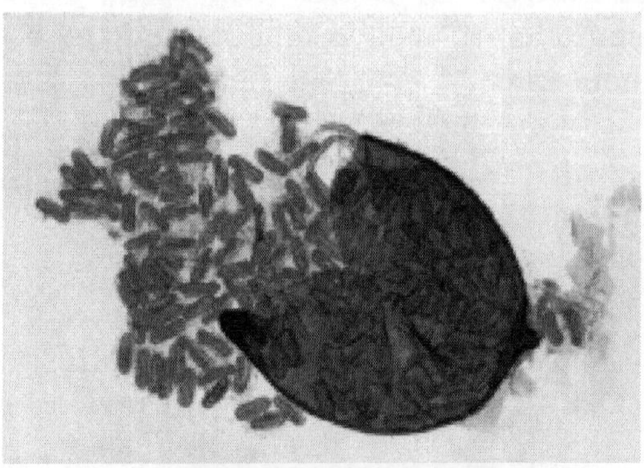

Figure: *Cyst nematodes retain most of their eggs within their bodies.*

Sedentary endoparasites are the nematodes for which host resistance has most often been identified. Relatively small differences in heritable characteristics within some plant species seem to determine whether or not a specific nematode will be able to establish giant cells in their tissues. Plants in which giant cells cannot be established, or in which they degenerate before a nematode can complete its life cycle, are resistant to that nematode. The attempted infection by the nematode may injure the plant, but the nematode cannot complete its life cycle in it.

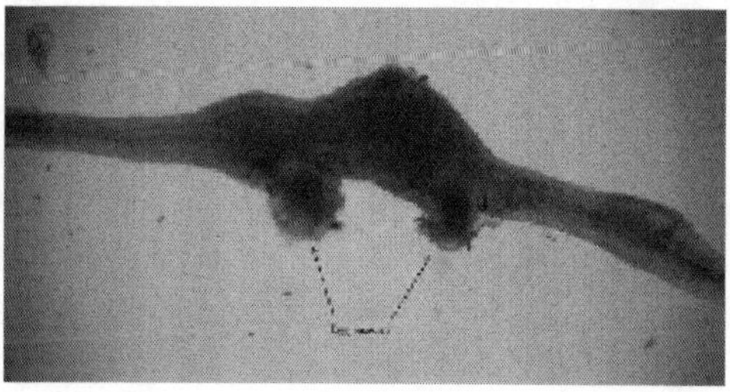

***Figure:** Root-knot nematode egg masses on a root surface.*

Diagnosing Nematode Problems

It can be very difficult to decide if nematodes are causing or are likely to cause significant crop injury. If a particular nematode pest was previously found in a site, it probably is still present. Plan to continue to take steps to manage it. In a location for which a complete history is lacking, the identities and population densities of nematode pests, hence the severity of damage which might be expected, can usually be determined by laboratory assay of soil and/or plant samples. Another common diagnostic problem is determining the role of nematodes when established plants are making unsatisfactory growth. This task is often difficult because few nematodes cause distinctive diagnostic symptoms. A sound diagnosis should be based on as many as possible of symptoms above and below ground, field history, diagnostic nematicide tests, and laboratory assay of soil and/or plant samples.

Above-Ground Symptoms

These are rarely, if ever, sufficient evidence to diagnose a root nematode problem. However, they are important because possible nematode problems are almost always first noticed because of abnormal top growth. Certain kinds of symptoms are typical of nematode injury to roots, and should always make one consider nematodes as a possible cause for the inferior performance. They can also be used to help locate the most severely affected areas in the planting after the problem is diagnosed.

Since most plant nematodes affect root functions, most symptoms associated with them are the result of inadequate water supply or mineral nutrition to the tops: chlorosis (yellowing) or other abnormal coloration of foliage, stunted top growth, failure to respond normally to fertilizers, small or sparse foliage, a tendency to wilt more readily than healthy plants, and slower recovery from wilting.

Woody plants in advanced stages of decline incited by nematodes will have little or no new foliage when healthy plants have substantial flushes, and eventually exhibit dieback of progressively larger branches. "Melting out"or gradual decline is typical of nematode-injured turf and pasture. Plantings which are stunted by nematodes often have worse weed problems than areas without nematode injury, because the crop is less able to compete with weeds than it should be.

Figure: *Yellowing of pepper plants resulting from nematode damage to roots.*

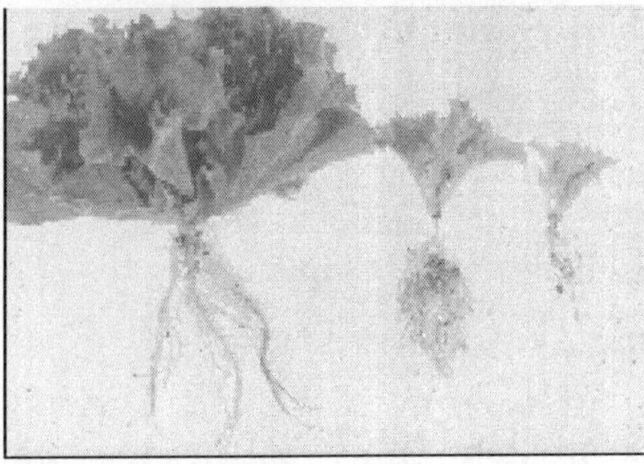

Figure: *Healthy lettuce (left) and stunted lettuce that has been damaged by nematodes. Healthy lettuce (left) and stunted lettuce that has been damaged by nematodes.*

Figure: *A bottlebrush being damaged by root-knot nematodes. Notice the thin canopy compared to the healthy plants behind it.*

The distribution of nematodes within any site is very irregular. Therefore the shape, size, and distribution of areas showing the most severe effects of nematodes will be highly irregular within the field. Nematodes move very few feet per year on their own. In the undisturbed soil of groves, turf, and pastures, visible symptoms of nematode injury normally appear as round, oval, or irregular areas which gradually increase in size year by year. In cultivated land, nematode-injured spots are often elongated in the direction of cultivation, because nematodes are moved by machinery. Erosion, land levelling, and any other force which moves masses of soil or plant parts can also spread a nematode infestation much more rapidly than it will go by itself. Nematode damage is often seen first and most pronounced in areas under special stresses, such as heavy traffic, excessive drainage because of slope or soil and dry areas outside regular irrigation patterns.

Below-Ground Symptoms

These may be more useful than top symptoms for diagnosing many nematode problems. Galls, abbreviated roots, necrotic lesions in the root cortex, and root rotting may all help in diagnosing nematode problems.

Root Galls and Other Root Swellings

The most distinctive nematode symptoms on roots are galls caused by *Meloidogyne* spp. They are small, individual, bead-like or fusiform swellings in some hosts. In other plants, galls may be massive lumps of fleshy tissue more than 1 inch in diameter, containing dozens of nematodes.

Some hosts, including many grasses, may not form any visible root swelling even though the nematodes successfully establish giant cells, mature, and reproduce. In such cases, an absence of galls does not necessarily prove that there are no root-knot nematodes present or that the plant species in question is not a host for that root-knot nematode.

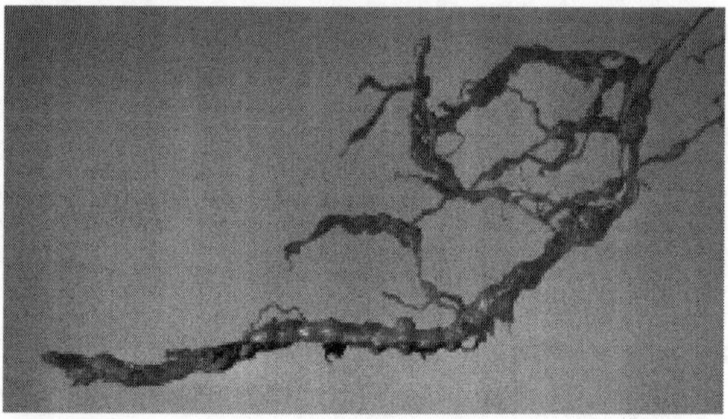

Figure: *Bead-like galls on cotton roots result from root-knot nematode infection.*

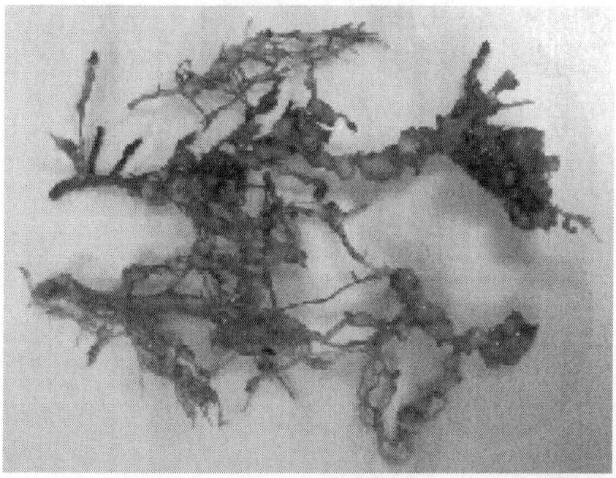

Figure: *Massive lumpy galls caused by root-knot nematodes on passion flower roots.*

Other root swellings must not be mistaken for root knot galls. Some ectoparasitic nematodes may cause root-tips to swell; by the time root-knot galling is visible, at least some of the galls are found distributed along the roots, away from the tip. Nitrogen-fixing bacteria cause swellings on the roots of most legumes (such as clovers, peas and beans).

These swellings, called nodules, are easily distinguished from root-knot galls by differences in how they are attached to the root and their contents. Nitrogen nodules are loosely attached to the root, and can generally be very easily removed; root-knot galls originate from infection at the centre of root, so they are an integral part of the root whose removal requires tearing the cortex apart.

In addition, fresh nodules should have a milky pink to brown liquid inside them, while root-knot galls have firmer tissues and contain female root-knot nematodes (creamy white beads less than l/32 inch in diameter) inside the gall tissues, near the fibrous vascular tissues of the root.

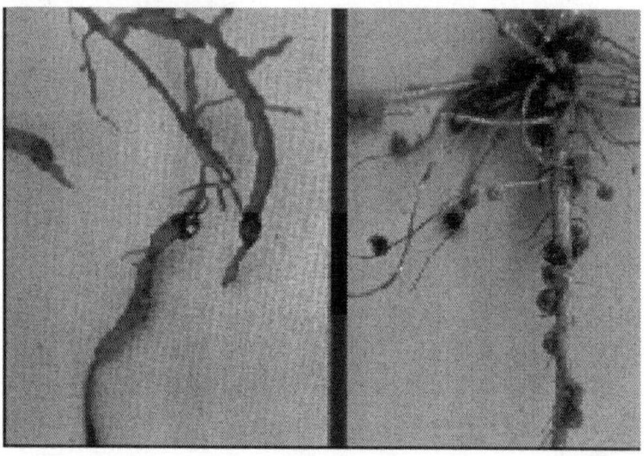

Figure: *Root-knot nematode galls (left) are an integral part of the root and cannot be easily removed. Nitrogen-fixing nodules (right) are attached to the root and are easily removed.*

Abbreviated Root Systems

Abbreviated root systems may be caused by several kinds of nematodes. Many ectoparasites which feed on root-tips, such as sting, awl, dagger, and stubby-root nematodes, cause root elongation to stop. These root tips sometimes swell to greater diameters than usual when they stop growing in length, and often become much darker in colour than uninjured root tips. Lateral roots often emerge from the root a short distance behind the injured main root tip; if a series of lateral roots are injured as they emerge, the end of the root may acquire a bunchy or bushy arrangement of very short roots that is very characteristic of sting or awl nematode injury to grasses and some other plants.

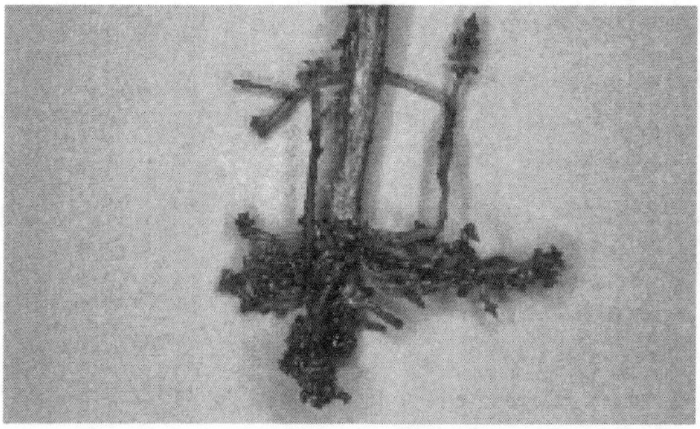

Figure: *Abbreviated "stubby-root" symptoms resulting from damage caused by ectoparasitic nematodes.*

Migratory endoparasites can also cause abbreviated roots. When lesion, burrowing, or lance nematodes injure the fleshy cortex of host roots, fungi which ordinarily could not penetrate the intact root are often able to colonize the injured tissues and, from there, infect the entire root. Rotted mature tissues at the tip of the root are a clue that endoparasitic nematodes and/or root-rot fungi, rather than ectoparasitic nematodes, may have shortened the roots.

Root Rot

Rot of large roots and storage organs (potatoes, etc.) is sometimes caused by nematodes alone. The potato rot nematode, *Ditylenchus destructor*, and yam nematode, *Scutellonema bradys,* are two examples. Burrowing nematodes often cause extensive necrosis, which looks much like dry rot, of fleshy tissues of large roots such as those of bananas and many tropical foliage plants.

Cyst Nematodes

An experienced observer can often see cyst nematodes on the roots of their hosts without magnification. The young adult females are visible as tiny white beads, about the size of a period on this page. After a female cyst nematode dies, her white body wall is tanned to a tough brown capsule containing several hundred eggs. Important cyst nematodes found in Florida include soybean cyst nematode on soybeans and a few leguminous weeds, beet cyst nematode (*H. schachtii*) on cabbage and related plants, St. Augustine grass cyst nematode on St. Augustine grass, and cactus cyst nematode on Christmas cactus and related plants. *H. cyperi* is a cyst nematode occasionally found infesting nutsedges (*Cyperus* spp.).

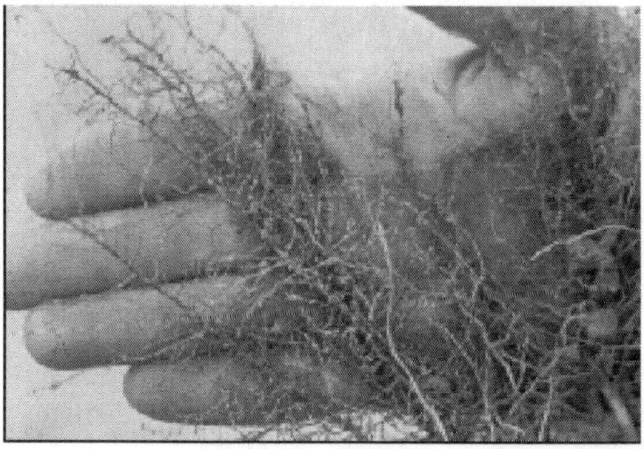

Figure: *The tiny white spots on these soybean roots are female cyst nematodes.*

Field History

Field history can provide valuable clues to the identity of nematode and other pest problems. A nematode which has been present in the field in recent years is probably there yet, and is likely to cause injury to susceptible crops if environmental conditions are favourable. Even without specific knowledge of pests in the field, records may help. If production records show a gradual decline in yields over a period of years, despite no change in cultural practices, nematodes may be indicated. A nematode infestation in a new field usually begins in a small area. It gradually intensifies in the original spot and is spread through the field by cultivation, harvest, erosion, and other factors which spread infested soil or plant parts. Therefore, the total effect of a recently introduced nematode is a gradual production decline for the field, as the percentage of the field that is involved and the severity of damage at any given area in the field increase over the years.

Diagnostic Nematicide Trials

Nematicide treatment may be used to help diagnose the role of nematodes in some situations. Laboratory determination of the kinds and numbers of nematodes present is sometimes not sufficient to determine the role of nematodes in a growth problem, or to predict injury to a crop to be planted. Application of a nematicide which is known to be effective against that pest is sometimes the most clearcut way to determine their importance, or at least the ability of an available treatment to remedy the situation. In such trials, it is best to apply the treatment to at least four areas of the suspect field, leaving adequate adjacent untreated areas for comparison. The nematicide product must, of course, be legal for application to the site and in the manner needed for the test.

Nematode Assay

A nematode assay is often necessary to complete a diagnosis. A variety of methods may be used to extract nematodes from soil and plant tissues. When they are identified and their relative population densities are known, the amount of risk to the crop from plant-parasitic nematodes can be determined. These risk levels have been assigned by nematologists at the University of Florida and represent the risk of economic loss to that crop. Such risk levels are determined through long-term experience of these nematologists with nematodes and with the crop in both growers' operations and in controlled experiments.

There are many crop/nematode combinations for which we do not have adequate research data to establish accurate risk levels. In these cases, the most important function of a nematode assay is to determine the kinds of nematodes present. For severe pests on high-value crops, the presence of barely detectable populations is often sufficient to justify use of an appropriate control measure such as a resistant cultivar, a nematicide, or a change of nematicide to one more effective for the particular nematodes present.

Principles of Nematode Management

Dozens of kinds of nematodes can injure crops; one or more of them occur in most soils in Florida. Although some can cause significant losses when present in low numbers, most do not cause economically significant damage unless their numbers are unusually high or the plant is also subject to unusual levels of stress caused by other factors. Carefully integrating as many as possible of the following components into the management program will help keep most nematode pests below damaging levels, and may simplify the decisions that must be made about selecting and applying nematicides when they are needed.

Prevention

Preventing a nematode problem is far better than trying to treat one after it is established. Many serious nematode pests are widespread, but a few are quite limited in distribution. For instance, the soybean cyst nematode is presently known in Florida only in the Panhandle and a few central and northern counties. Other nematodes which occur throughout the state may not occur in every field in a given area, and it is very worthwhile to avoid bringing any pest into a previously uninfested field. For example, if peanut root-knot nematode is in one field but not in a neighbouring field, one can avoid carrying that serious pest problem into the uninfested field by knowing that nematodes are spread in contaminated soil and plant parts.

Quarantine refers to a governmental regulatory action taken to prevent importing a pest into a previously uninfested area. It is an attempt to use official regulations to enforce the common sense rules that a grower should apply on his own to prevent spread of pests which have limited distribution.

Since most nematodes enter new areas or spread within an area by movement of soil or infested plants, most quarantine regulations and voluntary sanitation measures are intended to control the movement of contaminated soil and plant material. For instance, soybean cyst nematode can be spread in bits of soil carried with soybean seed from infested land. Federal quarantine regulations in effect for many years prohibited using seed grown in contaminated areas unless it had been thoroughly cleaned, to avoid importing soybean cyst nematode from those areas.

Good sense dictates working areas that are not infested with nematodes before moving to those that are infested, to avoid carrying contaminated soil or plants on his machinery to the uninfested field. Ornamental cuttings to be rooted should be taken only from uninfested plants or portions of plants from above ground which have never been rooted in a potentially contaminated soil. This prevents propagating populations of root-knot and burrowing nematodes that might seriously reduce growth and would cause the plants to be unfit for shipment to many potential markets because of quarantines.

Crop Rotation

Rotation is a very old practice for reducing soilborne problems. Many nematodes, soilborne disease organisms, and insects can reproduce and survive on only a few plants. Repeatedly planting a field with the same crop without interruption will enable any organisms which reproduce successfully on that crop to continue to increase. Rotation to non-host crops may cause many of those pests to cease reproduction and allow natural mortality factors to reduce their numbers. By carefully planning the sequence of crops to be planted in a particular field, it is often possible to avoid excessive build-up of pests that are important to all of the major cash crops in the cycle.

In a few instances, it is even possible to include a crop in the rotation that will help to control pests that have built up in preceding crops in the cycle. For instance, hairy indigo planted as a summer cover crop between spring harvest and fall planting of vegetables has been found to sharply reduce the populations of sting and root-knot nematodes in the Sanford area of central Florida. When managed properly as a summer cover crop, the hairy indigo helps reduce the populations of these two serious nematode pests, increases soil organic matter, and contributes some nitrogen to the soil, since hairy indigo is a legume. Similarly, pangola digitgrass has been used as a rotation crop to control burrowing and root-knot nematodes in vegetable lands in Florida and in the West Indies. However, pangola digitgrass is highly susceptible to sting nematodes, so is not a good choice for land infested with that pest.

The many kinds of nematodes that occur in Florida complicate selection of rotation crops, because crops which will reduce some species of nematodes may favour the increase of others. Despite the difficulty, a good rotation program should be a basic component of most land/crop management plans, because of the multiple benefits that can be derived from it.

Crop Root Destruction

A cultural practice which receives far less attention than it deserves is the immediate destruction of crop roots at the conclusion of the cropping season. It is a practice heavily emphasized in some areas of the country for high value crops such as tobacco. Fields should be tilled as soon as possible after harvest to kill all plants in the field and destroy their roots. Nematodes, soilborne diseases, and many soilborne insects will continue to feed and multiply on crop root systems as long as they remain alive.

In Florida, the prolonged period of high soil temperatures in the fall means that nematodes may have an additional 3-4 months after harvest of most crops before soil temperatures drop low enough that nematode reproduction is halted. When soil temperatures are high, each month that the root system continues to live represents an additional generation and potential increase of about 10-fold for root-knot nematodes and many others. Even in fall, when soil temperatures are gradually declining, a 2 month period represents at least one or two additional generations. Therefore, destruction of root systems as soon as the crop is finished will dramatically reduce the potential for increase of nematodes and other soilborne pests and should encourage their decline through normal mortality.

Flooding and Fallowing

In certain cases, flooding or fallowing may be used to help reduce numbers of nematode pests. Flooding has long been practiced for control of root-knot nematodes and some other soilborne pests in muck soils of the Zellwood and Belle Glade areas. It is practical only where the water level can be controlled easily and maintained at a high level for several weeks. It is also important to consider the possibility that flooding may actually spread some soilborne pests, if the water source is likely to contain disease-causing fungi such as Fusarium or nematodes such as soybean cyst nematode. Where flooding can be practiced, alternating periods of about 2 or 3 weeks of flooding, drying, and flooding again are apparently much more effective than a continuous period of flooding. The soil should be worked during the periods of drying to increase aeration and drying of soil and to prevent weed growth while the soil is exposed. Flooding probably kills nematodes by providing a long period without host plants rather than by some direct physical effect on the nematodes. Fallowing refers to leaving a field with no plants on it for a prolonged period. Most nematodes will decrease after a period of time without an adequate host plant. One study found that fallowing soil for 100 days during the winter lead to a 70% reduction in sting nematodes. Exposure to extremely high temperatures in the surface levels of the soil during the hot Florida summers may also reduce numbers of many kinds of nematodes. However, fallowing may be objectionable from a soil management viewpoint, because of the risk of losing soil organic matter, increased danger of erosion, and loss of productive time.

If fallowing is being used to help reduce nematode numbers, the field should be cultivated regularly to prevent growth of weeds and to expose new portions of the soil to the effects of drying and heating. If weeds are allowed to grow without control in a "fallow" plot, many kinds of nematodes may be able to survive and reproduce on the weeds, making the practice ineffective.

Resistance

When it is available, host plant resistance to a specific pest is usually the least expensive and most effective means of minimizing losses to that pest. However, successful use of varietal resistance requires knowing the extent of the resistance and which pests are present in a particular situation. There are "nematode resistant" varieties of tomatoes, soybeans, southern peas, sweet potatoes, cotton, and tobacco available for use in Florida, but each of these varieties has resistance which is effective against only one, two, or at most three, species of nematodes.

None are "nematode proof." Therefore, to be able to use the resistance available in a particular crop species, you must know the pest species present in the field. In addition, varieties with the appropriate resistance must be adapted to cultural conditions and requirements of your area. It still is necessary to use other methods to control any other nematodes that are present, because resistance against one or two species does not affect the ability of any other nematodes to injure the crop.

For example, if a farmer wishes to grow soybeans in a field which is infested with the sting, southern root-knot, and soybean cyst nematodes, he can take advantage of resistance to the root-knot and cyst nematodes in a variety such as 'Kirby' but still must use a nematicide to control sting nematode. Since the sting nematode is much more easily controlled with nematicides than are the other two pests, resistance to root-knot and cyst nematodes in Kirby simplifies the grower's chemical control program considerably. In another field where only the southern root-knot and cyst nematodes exist, no nematicide may be needed.

Biological Control

Nematologists in Florida and around the world are working very hard to identify and learn to manipulate natural enemies of nematodes so they can be used as biological control agents. Many different bacteria and fungi have been able to reduce populations of some kinds of nematodes under laboratory conditions, but successes at field scale are rare. Most organisms recognized to be promising for biological control of nematodes are quite specific in which nematodes they will attack, or have been very difficult to culture in sufficient quantities to be useful for field application, or both. The conditions under which each is most effective are often quite specific and limited. In all, then, commercially effective biological control as a means to reduce the effects of nematodes on lawns still appears to be many years away.

Nematicides

These are chemicals used to control plant parasitic nematodes. Two major groups of nematicides are distinguished by the manner in which they spread through the soil. Soil fumigants are gases in cylinders or liquids which spread as gases from the point where they are injected into the soil. Non-fumigant nematicides include a variety of water-soluble compounds which are applied to the soil as liquid or granular formulations.

Most belong to the carbamate or organophosphate families of pesticides. Their distribution in the soil depends on physical mixing during application and moving in solution in soil water. Some are absorbed by the plant and distributed to other parts of the plant ("systemic") where they may be effective for nematode or insect control.

There is no perfect nematicide for all purposes. Nematicides vary in their effectiveness against different kinds of nematodes, ease of handling, cost, effect on other classes of pests (weeds, disease organisms, insects), behaviour in different soils, toxicity to different plants, and availability.

There are many cases where some nematicides offer the only effective means of reducing or preventing nematode injury to a crop. There are others in which nematicides provide a quick and easy means of controlling a nematode problem which should not have occurred if the other available management options had been exercised. Sometimes, the practical and economic feasibility of using a nematicide depends on the level at which other management components are being maintained in the system. Generally, a higher yield potential and higher capital risk, because of such management components as fertility levels and irrigation, are more apt to justify the expenditure and effort to use a nematicide than are low levels of management or return. The factors affecting the decision to use a nematicide, and to select an appropriate one, are complex.

The succeeding sections which discuss the management options available for various crops grown in Florida include information about nematicides which are legally registered and recommended for specific purposes in this state. Since many pesticide registrations and growing conditions in Florida are substantially different from those in other areas, anyone from out-of-state should not use products or practices recommended in this Guide without consulting local authorities as to their legality and advisability. In all cases, the Product Label and supplemental labelling information must provide the legal guidelines for product use. All pesticide use must be consistent with the label.

Chapter 3

Nematode and Plant Disease Management

From a plant pathologist's perspective crop rotation can be one of the most effective and relatively inexpensive methods of managing a number of plant diseases. This practice has been advocated as a means of maintaining crop and soil productivity since the time of the ancient Romans, Greeks and Chinese. However, it is interesting to note that in a book on Roman farming, White (1970) indicated there was some suggestion from the historical literature that although Roman agronomists advocated crop rotation, this practice may not have been widely used by Roman farmers. The more things change the more they stay the same. As Karlen et al. (1994) states "One reason for farmer hesitancy to use crop rotation may be that agricultural scientists are still unable to explain the mysterious "rotation effect." Relative commodity prices, ease of production/marketing, and on-farm feed requirements are also important factors for producers when planning their crop rotations.

Although crop rotation is often advocated as a disease management tool, producers should have a clear understanding of the benefits of rotation so they can factor these potential benefits into their cropping decisions. Crop rotation can be thought of as a biological method of disease management. The success of this biological method of control is dependent on a number of factors including time, the environment, the nature of the pathogen you are trying to manage, and characteristics of the host crop. Rotation to a non-host(s) for a sufficient period allows enough time for decomposition of infested crop residues and/or a reduction in the viability of pathogen survival structures and the pathogen's ability to produce inoculum, thus eliminating a potential source of disease. As Cook & Veseth (1991) indicated in their book "Wheat Health Management" "...rotation allows time for natural enemies to destroy the pathogens of one crop while one or preferably two unrelated crops are grown." They also indicated that rotation acts like a natural type of "soil fumigation" where the collective activity of "antibiotic, predatory, and competitive organisms" help to eliminate plant pathogens from soil and infested crop residues.

Rotation represents a simple "meat and potatoes" approach to disease management. As a management practice it does not have the appeal of a new and improved crop variety and does not lend itself to the slick advertising strategies for many of the fungicides. In addition, depending on the commodity price situation it might even be considered a down right dirty word. Nevertheless, pathologists, agronomists, weed scientists, and soil scientists will keep flogging the practice of crop rotation as we enter 2000 and beyond. Why you might ask? Simply, it is because of the overall rotational benefits associated with crop and soil productivity, economic considerations, and the ability to manage weed and disease problems.

Considerations for Rotation to be an Effective Disease Management Tool

For rotation to be an effective strategy for disease management producers need to keep in mind that certain requirements or conditions must be met for this practice to be effective. Firstly, the disease source must be from the field itself or where there is limited potential for movement of plant pathogens from adjacent fields. Consequently, crop rotation will tend to be most effective for those diseases that are soil- or residue-borne. Crop rotation will not be effective for diseases where seed-borne pathogens are an important source of disease, such as the cereal smuts. Furthermore, rotation will not be effective where pathogens are mobile and readily move from one field to another or over long distances.

Secondly, the host range of the pathogen needs to be narrow. Crop rotation may not be an effective strategy for a disease like sclerotinia stem rot of canola because the pathogen that causes this disease has an extensive host range including most common broad-leafed weeds and crops, and thus would have ample opportunity to persist over growing seasons via other potential host plants. The final condition for rotation to be effective would be that pathogens are unable to survive for long periods in the absence of a suitable host. Thus, there should be a rapid reduction in pathogen viability and its capacity to produce infective structures, i.e. spores, within a relatively short period of time. Rotation will not be as effective for those pathogens that have the capability of producing long-lived resting structures or where they can survive in infested plant tissues that resist decay. *Sclerotinia sclerotiorum* produces resting structures known as sclerotia that are fairly long-lived. The blackleg fungus can survive in old canola root pieces for extended periods of time since these plant tissues are fairly resistant to decay and can persist for periods of three or more years.

Let us look at some examples where rotation can have a beneficial influence on disease management, where rotation may need to be combined with other strategies, and a bit of a spin on the traditional concept of crop rotation.

The Rotational Benefit of Pulse Crops

One feature of many ancient crop rotations was the inclusion of legumes including pea, bean, vetch, and lupins. The benefit of including legumes in crop rotations was recognized in ancient times and is also recognized today. Crop rotations that include legumes have the ability to provide succeeding crops with nitrogen, reduce disease incidence and improve weed control strategies. Inclusion of legumes in rotation will cause changes in soil fertility, soil microorganisms, soil organic matter, soil water and crop responses (Biederbeck et al. 1995). The focus of this section will be the rotational benefit of pulses with regard to root disease management in subsequent cereal crops.

Research conducted in Australia over the past twenty years has illustrated the benefit of pulse crops for subsequent cereal crops. Reeves et al. (1984) studied the effect of lupin on subsequent wheat crops from 1974 to 1979. Inclusion of lupin either in the preceding year or even two years previous, significantly reduced the level of take-all, caused by the fungus *Gaeumannomyces graminis* var. *tritici*, and increased yield compared with a continuous wheat rotation. Doyle et al. (1988) found that the incidence and severity of common root rot caused by the fungus *Cochliobolus sativus* was reduced in wheat following lupin compared with wheat following wheat. Yields were also significantly higher in wheat following lupin and this increase was attributed to suppression of disease and better nitrogen fertility.

Rotational benefits with regard to disease management have not been confined to lupin. Cotterill & Sivasithamparam (1988) also found a reduction in take-all after rotation to field pea, lupin or oats. Recently, Felton et al. (1998) demonstrated that crown root in wheat, caused by the fungus *Fusarium graminearum* Schwabe, was reduced and wheat yields increased in a chickpea-wheat versus a wheat-wheat rotation. Overall, mean incidence of crown rot was 12% for the chickpea-wheat rotation compared with 30% with the wheat-wheat rotation. Felton et al. (1998) attributed the increase in yield with wheat preceded by chickpea to enhanced nitrogen fertility and a reduction in disease. In western Canada, Stevenson & van Kessel (1996a) found a significant reduction in common root rot severity of wheat with a pea-wheat versus a wheat-wheat rotation at one of two sites in Saskatchewan. In addition, Stevenson & van Kessel (1996b) in a landscape-scale study found a slight reduction in common root rot severity with the pea-wheat rotation, but this reduction was not significant. Bailey et al. (1992), also found reduced levels of common root rot when spring wheat was preceded by field pea or fallow compared with spring or winter wheat.

Research conducted in the United Kingdom also illustrated the disease management benefit of legumes on subsequent barley crops. Dyke & Slope (1978) found that over a range of nitrogen rates (0 to 113 kg/ha) take-all was severe in barley after barley, but was a minor problem in barley after oats, beans or red clover. Yields were reduced in barley after barley compared with barley after oats, beans or red clover at all nitrogen rates.

At Fort Vermilion, Alberta we have been involved with a crop management field trial evaluating conventional versus zero tillage systems and legume-based crop rotations. The trial was originally established in 1992. Precrop treatments for this study included red clover as a green manure, field pea as a seed crop, summerfallow and continuous wheat. The crop rotations included: 1) field pea-wheat 1-canola-wheat 2 (FP/W1/C/W2), 2) red clover-wheat 1-canola-wheat 2 (RC/W1/C/W2), 3) fallow-wheat 1-canola-wheat 2 (F/W1/C/W2), and 4) wheat-wheat 1-wheat-wheat 2 (W/W1/W/W2). Common root rot (CRR) severity in the wheat was assessed in early August of 1994, 1995 and 1996. Overall, CRR severity was reduced under zero versus conventional tillage. Root rot severity was also reduced in rotations with field pea, red clover or canola as precrop treatments compared with continuous wheat. In addition, CRR severity was reduced in legume-based rotations compared with non-legume rotations.

Crop Rotation and Cereal Leaf Diseases

In a study of commercial barley fields in Alberta from 1995 to 1997, there was a tendency for lower levels of scald and net blotch to occur when barley was planted in fields previously cropped to a non-host. For scald, a rotation of 1-2 years away from a susceptible host maybe sufficient to allow for the destruction of most infected plant tissue, since the scald pathogen primarily infects leaves and leaf sheaths, which decompose fairly rapidly. However, a longer rotation will be needed for net blotch, since the fungal pathogen that causes this disease also infects barley stems, which are more resistant to decay and will persist, especially the node tissue, for a longer period of time. Results from the Alberta barley leaf disease study also indicated that extended rotations might not always reduce a producer's risk of scald and net blotch. Substantial levels of both scald and net blotch developed in fields not planted to barley or forage grasses for the previous four years. Average percent leaf area diseased (PLAD) of the flag leaf-1 in each year ranged from 0.9 to 3.2% and 3.5 to 9.9% for scald and net blotch, respectively. However, in some individual fields scald and net blotch levels on the flag leaf-1 exceeded 30 and 50% PLAD, respectively. Most of the high levels of disease, especially for net blotch occurred in fields of Harrington. One potential source of disease in these fields may have been seed-borne inoculum, especially for the causal agent of scald, which is a rain-splash dispersed pathogen with limited potential for field-to-field movement of disease. The causal agent of net blotch is a fungus that produces wind-borne spores and thus, will have more potential for field-to-field movement.

In Alberta, Skoropad (1959), Lee et al. (1999), and Xi and Burnett (1997) found that *Rhynchosporium secalis*, the causal agent of scald, infected barley awns, lemmas, paleas, and pericarps. In the 1950's, Skoropad (1959) found that seed samples collected randomly from Alberta barley crops had from 1 to 10% scald infected seed with an average of 2.0%. More recently, during the summer of 1996, seed from a regional variety trial located at Westlock, Alberta was found to have scald infection levels that ranged from 0 to 10% infected seed.

The movement of scald into scald-free areas via infected seed has been suggested by a number of researchers. Research in the 1950's and 1970's demonstrated transmission of *R. secalis* from infected seed to seedlings mainly under greenhouse conditions. Reed (1957) also observed that infected seed that remained on the soil surface after seeding could act as a source of inoculum. In some countries, including India, New Zealand and England seed-borne inoculum of net blotch is considered to be an important source of disease. At Lacombe Alberta, Piening (1968) studied seed to seedling transmission using seed that was covered with cultured net blotch mycelium or ground net blotch infected leaves. In the field, seedling infection levels were 0.5 to 1.5% and 0 to 0.15% for infested and control seed, respectively.

At Indian Head, Saskatchewan, Bailey et al. (1992) found that leaf spot severity was higher in wheat following spring or winter wheat compared with field peas or summerfallow. Pedersen & Hughes (1992) found in the Parkland area of Saskatchewan, that when environmental conditions were favourable for disease development Stagonospora/Septoria leaf spot (*Stagonospora nodorum*) severity was reduced in wheat by a rotation with two years between wheat crops, but not for rotations with only one year between wheat crops. However, when environmental conditions were less conducive to disease, a rotation with one year out of wheat reduced Stagonospora/Septoria leaf spot severity to the same extent as a rotation with two years between wheat crops. In southwestern Saskatchewan, Fernandez et al. (1998) found in a research trial and in commercial wheat fields that leaf spot severity was not always lower in wheat planted after a non-host compared with continuous wheat. They concluded that at least two years out of wheat were required for decomposition of infested residues to reduce wheat leaf spot levels. In 1996 and 1997, a wheat leaf disease survey was conducted in commercial fields in Alberta. In both years, similar leaf spot levels (septoria complex and tan spot) were found in wheat grown on wheat and wheat grown on a non-host. Furthermore, leaf spot levels in wheat fields cropped to non-hosts for the previous two to four years tended to be similar to wheat planted on wheat residue.

Research from the United States has indicated that seed infection in wheat may be a factor that needs to be considered when using rotation as a means of cereal leaf spot management. In Florida, Luke et al. (1983) found decreased development of Stagonospora/Septoria blotch, caused by *S. nodorum*, with either a 1 or 2 year rotation using uninfected seed, but not when *S. nodorum* infected seed was used. Similar results were also found by Milus & Chalkley (1997) in Arkansas where low (7%) and high (34-40%) levels of seed infection with *S. nodorum* led to significant leaf infection in an experiment planted in an area not cropped to wheat for at least the previous four years. They concluded that crop rotations should be combined with appropriate seed treatments to help prevent outbreaks of Stagonospora/Septoria blotch. In New York State, Shah et al. (1995) found significant relationships between *S. nodorum* seed infection levels of up to 40% and subsequent development of Stagonospora/Septoria blotch in winter wheat. Seed to seedling transmission of *Pyrenophora tritici-repentis*, the causal agent of tan spot of wheat, has been demonstrated under greenhouse conditions, and under laboratory conditions and in artificial potting media under outside conditions.

At present little evidence appears to exist, especially in wheat, concerning the potential role of seed-borne pathogens as a source of inoculum for cereal leaf spot epidemics in western Canada. Recent seed surveys have shown that over a three year period from 1995 to 1997 between 71 and 97% of barley grain samples in Alberta had detectable levels of infection with *Pyrenophora teres*, the causal agent of net blotch. Samples from 1995, 1996, and 1997 had overall average seed infection levels of 7.1, 12.1, and 22.6%, respectively. The maximum observed seed infection levels were 82, 81, and 89%, for 1995, 1996, and 1997, respectively. The frequency and level of infection tended to be higher for 2-row barley samples compared with 6-row samples. Based on results from Edney and Tipples (1997) the majority of 2-row samples would likely be the variety Harrington. The same survey showed that over a three year period from 1995 to 1997 between 39 and 75%, and 92 and 100% of wheat grain samples in Alberta had detectable levels of infection with *P. tritici-repentis*, and *S. nodorum*, respectively. Samples from 1995, 1996, and 1997 had overall average seed infection levels with *P. tritici-repentis* of 0.6, 2.0, and 2.1%, respectively, while yearly average infection levels with *S. nodorum* were 22.3, 16.9, and 16.7, respectively. The maximum observed seed infection levels with *P. tritici-repentis* were 3, 9, and 12%, for 1995, 1996, and 1997, respectively, while maximum levels of seed infection with *S. nodorum* were 43, 54, and 54%, respectively.

Barley Variety Rotation, A Spin on the Traditional Idea of Crop Rotation

Currently, a significant proportion of the diet utilized for beef production in Alberta is barley silage and feed grain. Alberta accounts for close to 50% of the total barley acreage in western Canada. Barley is an attractive feedstock as a result of its high yield, protein and digestibility and because farmers are familiar with its production, have ready access to seed and have a choice of seed or silage production. An estimated 48% of the barley grown in Alberta is for feed purposes and in some regions this can be as high as 60-70%. To ensure a constant supply of feed material many farmers will often grow barley continuously for several years.

For example, in a survey conducted by McLelland (1992) an average of 36% of the producers grew barley on barley with the proportion increasing to 75% in the western parts of central and southern Alberta. However, continuous barley production can result in significant disease problems, especially where the same variety is grown year after year. Where a continuous supply of feed grain is needed and potential alternatives are lacking, rotation to non-hosts as a management tool for leaf diseases of barley is not an option.

At Lacombe, we are looking at the potential of rotating barley varieties with different genes for resistance to diseases like scald as a means of reducing the impact of leaf spot diseases. Rotating barley varieties with different genes for resistance would be analogous to what producers are currently doing to prevent the build up of herbicide resistant weeds by rotating herbicides with different modes of action.

Different barley varieties may have different sources of disease resistance, especially if they are from different breeding programs. Growing varieties with different sources of resistance and perhaps combined with rotation to suitable non-hosts may help to prevent the selection and build-up of a specific pathogen race, and thus prolong the usefulness of a particular source of resistance.

In 1998, a 3-year experiment was set up at Agriculture and Agri-Food Canada, Lacombe to investigate the potential of barley variety rotation as a means of reducing the impact of leaf diseases and maintaining crop productivity. Preliminary results will be presented from the trial at Lacombe.

Summary

Studies in Australia, the United Kingdom, and western Canada have shown that including legumes in rotation had a beneficial effect by reducing the level of cereal root diseases. In the trial at Fort Vermilion, Alberta, including legumes in rotation decreased the incidence of root diseases compared to continuous wheat. In addition, zero tillage crop management practices reduced the level of disease compared to conventional tillage crop management practices. With legumes in rotation and using zero tillage, the level of CRR in wheat 1 was reduced by approximately 30%. Wheat yields for rotations with field pea, red clover or fallow were similar, and were higher than those observed under continuous wheat.

For some diseases, rotation to non-hosts for a sufficient period will allow enough time for decomposition of infested crop residues and/or a reduction in the viability of pathogen survival structures, thus eliminating a potential source of disease. However, this may not be the only mechanism by which rotation to pulse crops provides a non-nitrogen benefit to subsequent cereal crops. As Cook and Veseth (1991) have suggested "antagonism of root pathogens in soil and around roots can be maximized by the same practices that favour the conservation or elevation of the organic matter content of the soil." Rotations that include pulse crops, especially under conservation tillage may be a method of accomplishing this. Growing pulse crops and their subsequent impact on soil quality may favour increased population levels of various soil organisms and enhance their diversity and activity. These soil organisms may then colonize crop residues before pathogens or displace pathogens from infested residues. In addition, these organisms may produce a variety of substances or actively antagonize or parasitize soilborne plant pathogens. Ultimately, these activities would have the effect of reducing survival and inoculum production by pathogens present in the soil environment. Research conducted by Lupawayi et al. (1998) in the Fort Vermilion crop management study described previously, demonstrated increased soil microbial diversity in rotations with field pea or red clover compared with continuous wheat.

Crop rotation in itself may not be sufficient to provide effective control of all diseases that affect crops in western Canada. Inoculum for many diseases may result from a variety of sources and include not only infested residue, but also soil- and seed-borne sources. For these diseases, producers will need an integrated approach where rotation combined with variety resistance, seed quality, and fungicides would be used to provide more consistent disease management. For example, although rotation to non-hosts may help to lessen the risk of cereal leaf diseases, this practice may need to be combined with other strategies such as reduced levels of seed infection to provide consistent long-term management of these diseases. Although seed-borne inoculum may represent a relatively minor source of disease, cereal leaf spot diseases are polycyclic in nature and as a consequence have the potential to build up to damaging levels in a relatively short period of time. Under favourable conditions, the causal agents of these diseases can complete and repeat their life cycles every 7-14 days depending on the particular pathogen and prevailing weather conditions. Seed-borne inoculum may be a consideration for producers where intervals between cereal crops are more than 2-3 years, a highly susceptible variety is being grown, and where there is potential for favourable weather conditions to promote rapid disease development during the growing season. However, more research is needed to look at the potential role of seed-borne disease in the development of cereal leaf diseases in western Canada.

For producers to truly capitalize on the potential benefits of crop rotation for disease management, they need to have a good understanding of the life cycle of the disease they are trying to manage. This knowledge will help to indicate when rotation will or will not be effective. In addition, it will also indicate where rotation will need to be combined with other management tools in an integrated strategy for disease management. Producers should consider the use of a range of control strategies to provide more consistent long-term disease management. Field planning, general agronomics, variety, seed choice, seed quality, seed treatment, balanced fertility, field scouting, in-crop fungicide application, and harvest management are all tools that producers have as part of their "risk management toolbox" for plant diseases.

Nonchemical Disease Control

With the increasing concern about use and misuse of pesticides in commercial agriculture and home gardens, there are more and more inquiries for organically grown commodities every year. Nonchemical control practices for plant diseases have been known and recommended for years. The backbone of any integrated pest control program must always include cultural and sanitation practices, two important components of nonchemical disease control.

Unfortunately, disease problems may begin as soon as seeds are planted and can continue into harvest and storage. Plant diseases may be caused by several pathogenic organisms, such as fungi, bacteria, viruses, mycoplasmas and nematodes. In addition, nonliving factors, such as deficiencies or excesses of water, light, temperature, air pollution, pesticides and nutrients, can either predispose a plant to disease or directly cause plant injury.

Fortunately, many disease problems can be prevented or controlled without pesticides. Effective plant disease control begins at the onset of disease or even before symptoms appear.

Resistance

Effective disease control through resistance (a plant's tolerance or immunity to a disease) is based on knowledge of the diseases that occur in an area. Always choose varieties of plants that are adapted to Colorado growing conditions. Many vegetable, fruit and ornamental plant varieties are available with resistance to one or more diseases.

For example, when purchasing tomato varieties, always select plants labelled "VFN," "VFNA," "VFNT," etc. This indicates that the plants are resistant to Verticillium wilt (V), Fusarium wilt (F), southern root-knot nematode (N), early blight (A), or tobacco (tomato) mosaic virus (T). Selecting resistant plants may eliminate many disease problems. Contact your Colorado State University Extension county office for lists of plant varieties that are successfully grown in this area.

Exclusion

Exclusion is preventing the entrance and establishment of disease-causing organisms (pathogens) into areas where plants are grown. This means avoid bringing diseases into the garden or moving them around within the garden.

Use certified, disease-free seed or transplants. Examine the leaves and root systems of transplants and eliminate or destroy diseased plants. Either raise your own transplants in sterilized beds or buy them from a reputable dealer. Do not purchase transplants with galls or swellings on their roots or plants that have a brown discoloration on the stem at the ground line. Galls or swellings may indicate root-knot nematode infection. A brown stem discoloration may mean the presence of damping-off organisms. Also, avoid transporting soil or tools from known disease areas to disease-free areas.

Eradication

Eradication is the elimination of the disease-causing organism after it has become established on a plant. Eradication can be accomplished by several methods.

Sanitation

Plant pathogens are less likely to survive if organic matter is quickly decomposed. Remove plant debris or infected plant parts after each growing season. Turn the soil after harvest to help break down small roots that may harbour nematodes, fungi or bacteria. Gardeners may compost dead plants if they have a good composting system; otherwise, these piles may serve as a source of pathogens.

Prune or remove twigs and branches of woody plants affected with fire blight and other bacterial or fungal canker diseases.

Keep gardens weed free. Weeds often are another source of pathogens. Eradicate weeds to break the life cycle of pathogens and control them. Weed removal also can increase air movement and thus decrease conditions that favour disease development.

So that pathogens do not spread from one area to another, always disinfest machinery and other tools with steam, hot water under pressure, or a 10 percent solution of household bleach diluted with water.

Crop Rotation

Avoid planting the same crop in the same area of the garden year after year. Continuous culture of the same kind of crop provides an opportunity for pathogen buildup.

For example, rotate leafy vegetables with grains or corn, or rotate annuals or biennials in seed and flower beds. It is best to grow the same or closely related plants in the same soil only once every three to five years. This practice starves out most pathogens that cause leaf, flower and stem diseases.

Crop rotation is not as effective against soilborne organisms, those fungi, bacteria and nematodes that persist in the soil for up to 10 years or more.

Soil Sanitation Treatments

Occasionally, disease-causing organisms that live in the soil may build up and prevent satisfactory growth of plants. Pathogen-free soil is desirable for houseplants, transplants and garden plots. Sterile potting mixes are available at many garden centres. However, it may be desirable to sanitize small quantities of soil on your own.

There are several nonchemical methods available to eradicate or reduce pathogens in the soil. The use of dry, steam or solar heat are the most effective nonchemical means to sanitize soil. The time to treat soil is before seeding or transplanting. Soil to be treated must be easily crumbled and without clods or large pieces of plant debris. Add any amendments (manure, compost, peat moss, etc.) before treatment. Soil also must have proper moisture. To test for this, gently squeeze a handful of soil. When the hand is opened, the soil ball should break apart somewhat. If it doesn't and the ball cannot be broken apart by gently pushing down on top of the ball, the soil is too wet. Oven sterilization. Place soil evenly but not more than 4 inches deep in nonplastic containers, such as seed flats, clay pots, and glass or metal baking pans. Tightly cover each container with aluminium foil. Insert a meat or candy thermometer through the foil into the centre of the soil. Set the oven temperature between 180 and 200 degrees F.

Heat the soil to at least 180 degrees and allow it to remain at this temperature for 30 minutes. Do not allow the temperature to go above 200 degrees, because this may cause products that are toxic to plants. After heating, cool and remove containers from the oven. Leave the aluminium foil in place until you are ready to use the soil. The heated soil will give off an odor. Microwaves or outdoor cookers also can be used.

Pressure-cooker sterilization. Pour several cups of water into the cooker. Place no more than 4 inches of soil in shallow containers on a rack out of the water. Level the soil, but do not pack it down. Cover each container with aluminium foil. Stack the containers to allow steam circulation. Close the lid, but leave the steam valve open until all the air is forced out and steam begins to escape. Then close the steam valve and heat at 10 pounds pressure for 15 minutes. Turn off the heat, allow the containers to cool, and remove. Leave the aluminium foil in place until you are ready to use the soil.

Steam sterilization without pressure. Pour about 1 inch of water into the sterilizing container. Follow the soil preparation procedures listed earlier. Place filled soil containers on a rack that will hold them out of the water. Close the lid and bring the water to a boil. Open the lid just enough to prevent pressure from building up. When the steam begins to escape, continue boiling for 30 minutes. Turn off the heat and replace the lid. Remove the soil when cool.

Soil solarization. This method uses the sun's energy to heat small areas of soil to temperatures that are lethal to many soilborne organisms and weed seeds. Treat during the summer when there are high air temperatures and intense solar radiation. With a tiller or a shovel, loosen the soil to be treated, wet it and cover it with a thin, clear, polyethylene (plastic) film. Seal the edges of the plastic sheet with soil to prevent heat loss and retain moisture. Leave the plastic in place for several weeks. The longer the soil is exposed to the heat generated by the solarization process, the greater the kill of undesirable organisms. Avoid toxicity from heated soil. With heavier soils and soils that contain a large amount of organic matter (manure, compost or peat moss), a toxic effect from heat sanitation may occur. This can cause poor seed germination, plant growth abnormalities or plant death. The toxicity is caused by an accumulation of ammonium compounds, soluble organic compounds, minerals or salts when the soil is heated too long or at too high a temperature. If soil toxicity is a problem, heavy irrigation of the treated soil may leach out many of these substances. Storing the soil two to three weeks without a cover also reduces soil toxicity.

Cultural Management

Cultural management involves avoiding the onset of disease. To accomplish this without pesticides, create an environment unfavourable to pathogens.

- Don't work in the garden when plants and soil are wet. Spores and cells of disease-causing organisms can spread from one plant to another and initiate new disease. Wet soils are easily compacted, which can decrease the amount of oxygen in the soil.
- Make sure plants are spaced properly. Air movement decreases when plants are grown too close together. This allows moisture to remain on leaves for longer periods of time. Wider spacing in beds and landscape plantings promotes rapid drying after wet periods and stops development of foliage, flower and fruit pathogens.
- Avoid excessive soil moisture. Overwatering enhances seed decay, damping-off and root rot diseases. Try not to plant in areas that have poor drainage or where water stands for several days following rains.

- Fertilize plants properly based on soil nutrient analyses using either organic or commercially prepared (inorganic) fertilizers.

In other words, use good cultural management. Healthy plants are less likely to have disease problems than weak, undernourished ones. Grow plants under optimum conditions and there will be fewer disease problems. Control of most plant diseases can be accomplished without pesticides. Use sound cultural practices, sanitation and well-adapted plant varieties to reduce disease problems. It is important to realize that you must accept some disease loss. Don't expect a perfect garden or plant if you do not want to use chemicals.

Disease Control in the Home Vegetable Garden

Plant diseases can be a significant problem in home gardens. Most vegetables are susceptible to a number of diseases. Wilts, leaf spots, blights and fruit rots are just a few of the disease problems that plague vegetables gardens every year.

Plant diseases are caused by four primary types of organisms: fungi, bacteria, nematodes and viruses. These organisms are often referred to as *plant parasites* or *pathogens*. Plant diseases caused by fungi and bacteria are most common when rain showers and/or heavy dews are frequent and temperatures are warm. During these times, scout the garden regularly for disease. Viral diseases can occur at any time, and nematode damage can be more evident if conditions are dry. Home gardeners can reduce the occurrence of many diseases with sound cultural practices.

Site Selection

Although few people have an ideal garden spot, site selection is important for a successful garden. A home garden site should be well drained. Avoid wet and poorly drained soils. Excessive soil moisture will favour root and crown diseases such as damping-off, crown and root rots caused by fungal soilborne pathogens. Most vegetables require full sunlight for maximum yield. Full sun speeds drying of the foliage, which reduces the incidence of most foliar diseases.

Crop Rotation

Crop rotation is very important in reducing losses to vegetable diseases. Continuous plantings within the same plant family of vegetables provide opportunities for pathogen buildup. Do not plant the same family of vegetable in the same areas year after year. Grow the same plants or closely related plants in the same soil only once every 3 to 5 years. This practice starves out most pathogens that cause stem and leaf diseases.

Unfortunately, crop rotation does not work against soilborne problems such as root and crown diseases caused by the fungi *Phytophthora, Rhizoctonia, Pythium* and *Sclerotium,* and vascular wilts caused by the bacteria *Pseudomonas.* These organisms are long-lived in the soil and affect many plant families. You can get only a limited benefit against *Fusarium* wilt.

Disease Free Seed and Transplants

Many plant diseases can be seed-borne. Do not save seeds from year to year. This is important to prevent a number of diseases, including halo blight, common blight, anthracnose of snap beans, and bacterial spot of tomatoes. Commercially available seed is often produced in the arid western United States where seed-borne pathogens do not occur and/or cannot survive. Seed of some vegetable varieties, however, are not produced in the west, and local seed must be used. Purchase seed from a reputable dealer because you cannot distinguish healthy seed from diseased seed.

Seed are protected by fungicides often applied to the seed by the company. Seed can also be treated by the home gardener. These treatments give young seedlings some protection from soilborne pathogens as they germinate and emerge. They do not, however, control pathogens that may attack the plant after the seedling stage. Some pathogens are carried on the surface of seed and can be controlled with fungicide treatment. Fungicides are not effective on pathogens carried beneath the seed coat. Gardeners starting a crop from transplants should examine transplants carefully before purchasing. Select healthy transplants that appear disease and insect free.

Use Resistant Varieties

Using resistant varieties is the most efficient way of controlling vegetable diseases. Make an effort to buy resistant varieties when they are available. Seed catalogues generally list resistant traits of different vegetable varieties. For example, disease resistant tomatoes are designated "VFN." These letter indicate resistance to Verticillium wilt (not common in Georgia), *Fusarium* wilt and nematodes. In tomatoes, there is also resistance to *Stemphyllium* (gray leaf spot) = S, Tobacco Mosaic Virus = TMV, *Alternaria alternata* stem canker = A or ASC, *Septoria* leaf spot = L and Tomato Spotted Wilt = TSW.

Other vegetables such as squash, pepper, beans and peas also have disease and nematode resistance. Note that plant disease and nematode resistance are relative terms, as resistant varieties have differing levels of resistance. Resistance does not always mean the plants are immune to these problems. Some seed catalogues may indicate tolerance to a certain pest. Tolerance usually indicates that a particular variety of vegetable may produce fruit of acceptable yield and quality while exhibiting moderate to high levels of disease.

Planting Date Management

Planting dates can be an effective tool for disease management. Follow the recommended planting dates for the particular vegetable grown. For example, okra should be planted when soil temperatures are warm for good germination and growth. Planting this crop when soil temperatures are cool can cause increased incidence of soilborne diseases such as soreshin and damping-off. Using transplants may help avoid early disease problems so long as the transplants are from a reputable source and are disease free. Plant crops such as corn and beans earlier, because their seeds germinate in cooler soils. It is also advantageous to plant these particular crops early to escape severe virus infections. This is particularly true with squash. Aphids, which transmit many virus diseases, are at lower population levels early in the season. High aphid populations late in the season can lead to more virus problems.

Trap Crops

Using trap crops may sometimes help manage virus diseases and aphid populations. A few rows of a trap crop around the vegetable garden (such as rye or corn) will cause aphids to feed there first, possibly loosening the virus they may carry. Trap crops can limit aphid damage and help reduce the incidence of virus diseases, which are sometimes severe on beans, cucurbits and solanaceous crops.

Proper Spacing and Trellising

Proper spacing and trellising can reduce the occurrence of vegetable diseases. Space plants properly to allow growth and air circulation. Humid and wet conditions occur if plants are crowded and unable to dry quickly. Extended periods of wet foliage are ideal for development of fungal and bacterial diseases. Crowded plants take longer to dry and provide favourable disease conditions. Crop protectants also cannot penetrate dense foliage, thus limiting their effectiveness. Staking or trellising tomatoes, pole of half runner beans, and cucumbers will prevent soil contact with the foliage and fruit. Air circulation will be better if these plants are trellised, promoting foliage drying and reducing disease severity. Limiting soil-plant contact reduces diseases such as fruit rots. Crop protectants also can be more effectively applied to trellised plants.

Proper Watering Practices

Avoid excessive soil moisture. Over-watering enhances seed decay, damping-off, and root and crown rot diseases. *Choanephora* fruit rot, which causes the whisker-like blossom end rot, can be increased by over-watering. If possible, irrigate by running water between the rows. Avoid splashing plants with soil, since any potential pathogens are soilborne. Drip irrigation is also a good method of watering without wetting foliage. Drip irrigation uses either a hose or tape that allows water to be applied slowly over time at the base of the plant or just beneath the soil surface. Garden drip irrigation kits are readily available. Avoid wetting the foliage when watering. Wet foliage is favourable for development of most foliar fungal and bacterial diseases. If wetting the foliage is unavoidable, irrigate in the morning so foliage can completely dry before evening.

Maintain uniform soil moisture. This can reduce problems such as blossom end rot (calcium deficiency) of tomatoes and peppers. Don't work in the garden when plants and soil are wet. Bacterial and fungal diseases spread easily from one plant to another by hands and clothing when above-ground plant parts are wet.

Use a Mulch Layer

Use a mulch layer of straw, bark, shredded paper or plastic to prevent soil from splashing onto plants and to prevent fruit from touching the bare ground. This will help prevent rots on mature fruit such as strawberries, tomatoes, squash, cucumbers and melons. Mulches also are a sound cultural practice to help conserve soil moisture and reduce weed infestations.

Proper Fertilization

Adequate and proper fertilization helps prevent vegetable diseases. Test soil three to six months before the growing season, and follow the recommendations to supply appropriate nutrient requirements and adjust soil pH. Avoid excessive amounts of nitrate forms of nitrogen, which encourage root rot diseases. Use ammonium forms instead (no more than 50 percent). Be sure the soil pH level is in the proper range for a particular crop. Proper pH prevents blossom end rot and encourages healthy growth of tomatoes and peppers.

Weed Free Garden

Weeds in the same families as some of the vegetable crops can be another source of diseases and insects. For example, some weeds can serve as virus reservoirs for several insect-transmitted viruses that can infect homegrown vegetables. Good weed control will increase air movement in the garden and decrease conditions — such as excessive moisture — that favour disease development.

Avoid Tobacco when Working in the Garden

If you use tobacco, wash your hands thoroughly before handling plants. This practice will prevent the spread of tobacco mosaic virus, which can infect many different kinds of vegetables, particularly solanaceous crops such as tomatoes and peppers.

Nematode Control

Nematode control is important in growing vegetable crops. Nematodes are microscopic soil-inhabiting roundworms that injure vegetables and other plants by feeding on their root systems, causing decay and/or galling. You can sample soil for nematodes by submitting a sample through your county extension office to the Extension Nematology Laboratory. The most damaging nematode in Georgia is the root-knot nematode, named for the knots it produces on infested roots. Soil solarization can be used to reduce the number of nematodes present. This involves tilling the garden and then covering the area with a clear plastic tarp for six to eight weeks. Solar radiation is trapped in the soil and can raise soil temperature high enough to kill most nematodes in the upper 3 to 4 inches of soil. In Georgia, the best time to solarize soil is June through August, when temperatures are at their hottest. Soil solarization requires intense solar heat to be effective. Repeated roto-tilling can also desiccate and kill many nematodes, although repeated tilling may decrease overall soil organic matter.

Pesticide Use

Use of pesticides should be the last defence available to home gardeners once all other disease-control options have been exhausted. Home gardeners have few fungicides or bactericides available compared to commercial growers. Chlorothalonil (Bravo® or Daconil®), maneb, mancozeb, Terraclor® (PCNB), sulfur and copper products can be used on certain crops. These products provide disease suppression across a range of foliar diseases and are more effective if they are applied in a preventive manner at the very onset or before a disease outbreak occurs. Chlorothalonil, maneb and mancozeb generally provide broad-spectrum disease suppression of most fungi. Terraclor is used as a transplant soil drench to suppress Southern blight and Rhizoctonia damping-off. Copper products suppress fungi but primarily reduce losses to bacterial pathogens. Sulfur is suppressive to fungi and is especially effective against powdery mildews. As with any pesticide, read the label and follow the recommended precautions to ensure the chemical is applied in a safe, effective manner. Contact your county extension office for specific information about pesticides.

Chapter 4

Controlling Diseases in Plants

Diseases of vegetables grown in the home garden may reduce both the yield and quality of vegetables. Controlling such diseases often determines success or failure and adds immeasurably to the pleasure derived from the garden. Diseases of vegetables are caused by microorganisms, including fungi, bacteria (including mycoplasmas and spiroplasmas), viruses, and nematodes. These organisms are spread by wind-blown or water-splashed spores, insects, infected seeds or transplants, by the movement of infested soil, and by humans handling wet plants.

Vegetable diseases can be successfully controlled using an integrated disease control program. Such a program involves the use of crop rotation, sanitation, disease-resistant or tolerant varieties, disease-free seeds or transplants, sound horticultural practices, balanced soil fertility (based on soil tests), and proper and timely applications of pesticides. There is no single control practice that can control all of the diseases of any vegetable crop. Therefore, several measures must be used to achieve satisfactory control.

Disease Control Before the Garden is Planted

Choose and prepare the site: The site for the vegetable garden should be well-drained, have adequate sunlight (8 or 10 hours of direct sunshine per day), and good air circulation. The soil should have adequate soil fertility and the proper pH (5.5 to 7.0). Information on garden location, soil fertility, and soil pH can be found in University of Illinois Extension Circular 1331.

Midwest: If the area has produced garden vegetables for several years, sanitation and rotation practices will reduce the risk of disease-causing organisms that have survived from preceding crops. Removing or turning under dead and diseased plant material in the fall is important for disease control. Crop residues from healthy plants can be composted and returned to the garden. Disease-causing fungi and bacteria can survive composting unless the decomposition of the crop residue is complete, a situation that seldom occurs. Therefore, we suggest that no diseased plant parts be placed in the compost pile. Some disease organisms will even survive complete composting, including corn smut; clubroot of cabbage, broccoli, cauliflower and other crucifers; Verticillium wilt of potato, tomato, pepper, and eggplant; and root-knot nematodes. Turning (plowing or tilling) under crop residues in the fall helps prevent the over-wintering of many disease-causing organisms. Besides destroying the "winter home" of various organisms, this practice also helps control the insects that transmit certain diseases. Also, crop debris that is turned under improves soil tilth and makes spring garden work easier.

Crop rotation: Crop rotation is an important control measure since many disease-causing organisms only attack related plants in the same family. Unrelated plants do not serve as hosts on which these organisms can multiply. Therefore, if possible, avoid planting any vegetables within each of the following groups in the same location more than once every three or four years.

Cole crops–(cabbage family) broccoli, Brussels sprout, cabbage, cauliflower, Chinese cabbage, mustard greens, kohlrabi, radish, rutabaga, and turnip. Cucurbits–(cucumber family) cucumber, gourds, muskmelon, pumpkin, squash (all types), and watermelon.

Solanaceous crops–(tomato family) eggplant, husk tomato (ground cherry), potato, pepper, tomato. Chenopodiaceous crops–beet (red and sugar) and spinach.

Amaryllidaceae–(onion family) chives, leeks, onions, shallots, and garlic.

Legumes–pea, beans (all types). Umbelliferous crops–carrot, parsnip, celery. For example, cabbage, broccoli, radish, or turnips should not be planted in the same location for two successive years. Cabbage could be followed with beans, a vine crop such as cucumber, or with sweet corn. Diseases affecting beans, peas, sweet corn, carrot, lettuce, spinach, rhubarb, okra, edible soybeans, and similar plants are usually specific to only one of these crops, and generally will not infect others or members of the major vegetable crop groups.

Choose disease-resistant varieties: The use of well-adapted, disease-resistant varieties is the simplest and most efficient method of controlling many diseases. Lists of disease-resistant vegetables recommended for use by Illinois gardeners are given in Circular 1331 (Vegetable Gardening in the Midwest. Additional information on disease resistance is also given in the following paragraphs. A listing of varieties and their resistance can be found in Circular 1363, Midwest Vegetable Production Guide for Commercial Growers.

Start with disease-free seeds, plants, and planting materials. It is important to start with disease-free seeds or planting materials (bulbs, tubers, transplants, sets, and so on) to keep from introducing serious diseases into the garden. In general, gardeners are encouraged not to save their own seed but to purchase seed from reputable seed dealers. Hot-water soaks and fungicide seed treatments may be used to control the seedborne disease organisms that cause seed rots and damping-off. Since temperature controls and timing must be precise, home gardeners usually buy seed that has already been treated with hot water by the seed processor. Hot-water treatment of cabbage, broccoli, Brussels sprout, and cauliflower seed is particularly important. Many vegetable seeds are coated with a fungicide at the time of purchase, as evidenced by the coloured coating on the seed. Untreated seed can be treated by the home gardener by placing a small quantity (the size of one or two match heads) of captan or thiram in the seed packet and shaking the packet for a minute or two to thoroughly cover the seed. Excess protectant may be sifted out before planting.

Do not plant diseased material (for example, transplants, sets, bulbs, or tubers). All planting material should be healthy and free of yellowing and brown or black spots, and should not be stunted or show poor development. Only certified, disease-free potato tubers or sweet potato slips should be used. Examine transplants thoroughly for signs of leaf or stem disease. It never pays to buy and plant diseased transplants, no matter what the price. Damping-off diseases affecting home-grown vegetable transplants can be controlled by the use of diseasefree soil and fungicides.

Cultural Practices

Cultural practices such as controlling weeds, planting at the right time, planting at the proper depth and spacing, employing cages or stakes, and watering when the temperature is rising to promote a rapid drying of the foliage can help control many foliar diseases. Mulches help to control fruit rots and aid in control of blossom-end rot of tomato, pepper, and vine crops by maintaining a uniform supply of moisture in the soil. Weed control in and around the garden reduces the risk of attacks by viruses that overwinter in these plants. Do not work or harvest wet plants, since this can spread many diseases, particularly those caused by bacterial and fungal pathogens.

Disease Control During the Growing Season

Many diseases, such as the early blight disease of potato and tomato, occur each year despite all preplanting precautions. For such diseases, applications of fungicides and bactericides to the growing plants may be needed.

The best way to apply these materials is as sprays before the disease occurs. Often a wetting agent such as liquid detergent or soap (1/2 teaspoon in 1 gallon) is added to obtain more thorough wetting and coverage of the foliage. Apply sprays to the point of run-off on a 7- to 10-day schedule. This maintains a fresh or effective covering of fungicide and protects the new growth. Fungicides and bactericides currently recommended for use on vegetables is given in the Illinois Pest Control Handbook which is updated each year.

Principal Diseases of Vegetables and Control Measures

Asparagus

Rust: Grow resistant varieties Start spray applications after harvest and continue until mid-August on a 10-day schedule with an approved fungicide.

Crown and root rots: Maintain proper fertility and good soil drainage. Avoid over-cutting.

Beans

Seed decay and damping-off: Use captan- or thiram-treated seed. Plant in a warm, moist, well-prepared seedbed.

Bacterial blights: Do not save seed from infected plants. Purchase seed from a reputable seed dealer. Spray a copper-based bactericide such as copper sulfate at the first sign of disease, and continue on a 7- day scheduled until harvest. Do not work among wet plants.

Rust. Grow rust-resistant varieties: Apply fungicides at the first sign of disease and continue on a 7-day schedule.

Mosaics. Grow resistant varieties..

Beets

Seed decay and damping-off: Plant seed treated with thiram or captan.

Leaf spots: Spray with an approved fungicide on a 7-day schedule.

Carrots

Seed rot and damping-off: Plant seed treated with captan or thiram.

Leaf spots: Practice a 2- to 3-year crop rotation. Spray weekly with an approved fungicide, starting about mid-June.

Aster yellows: Destroy infected plants as soon as they appear. Apply insecticide sprays for leafhopper control. Spray before removing infected plants.

Cole Crops: (cabbage, broccoli, Brussels sprout, cauliflower, Chinese cabbage, kohlrabi, mustard greens, radish, rutabaga, and turnip).

Seed decay, damping-off, black rot, and blackleg: Plant hot-water-treated seed which has also been treated with captan or thiram. Several black-rot-resistant cabbage varieties are now available.

Leaf spots and blights: Practice 2- to 3-year crop rotation for broccoli, cabbage, cauliflower, and Brussels sprout. Apply an approved fungicide on a 7-day schedule.

Clubroot: Purchase disease-free transplants. PCNB (Terraclor) may be added to the transplant water.

Fusarium yellows: Grow only varieties that are resistant to yellows.

Cucurbits (Cucumber, gourds, muskmelon, pumpkin, squash, and watermelon).

Seed rot and damping-off: Plant seed treated with captan. Plant in a warm, well-prepared seed bed.

Angular leafspot: Practice a 2- to 3-year crop rotation. Spray with a fixed copper bactericide, such as COCS. Remove diseased plant material. Plant resistant varieties whenever possible.

Anthracnose, Alternaria leaf blight, downy mildew, blossom blights, fruit rots and spots, and gummy stem blight or black rot. Practice 2- to 3-year crop rotation. Apply an approved fungicide at weekly intervals, starting when the plants are in the two-leaf stage. Adequate nitrogen fertility is essential for the successful control of Alternaria leaf blight. Plant resistant varieties when available.

Bacterial wilt: Spray with an insecticide to control cucumber beetles as soon as the seedlings "crack" the soil before they emerge. Continue weekly spraying until the plants are in bloom. Spray in the evening to avoid killing bees.

Fusarium wilt: Grow only cantaloupe and watermelon varieties that are resistant to Fusarium wilt.

Powdery mildew; Spray with an approved fungicide at the first sign of disease, and weekly thereafter.Grow resistant varieties.

Mosaic and other virus diseases: Grow resistant varieties where possible. Only cucumber-mosaicresistant cucumbers should be grown.

Eggplant

Seed rot, damping-off, and Phomopsis blight. Plant seed treated with captan or thiram.

Phomopsis blight and other fruit rots. Practice a 2- to 3-year crop rotation. Spray weekly with an approved fungicide when fruits are half size, or when disease first appears.

Verticillium wilt. Mulching with black plastic may help reduce losses.

Onion

Smut, seed rot, and damping-off. Plant seed treated with thiram or captan.

Leaf diseases: Practice a 2- to 3-year crop rotation. Spray weekly with an approved fungicide. Carefully follow label directions as regards the days between the last spray and harvest.

Bulb rots: Control leaf diseases. Also follow these recommendations: avoid storing improperly cured or injured bulbs; let all onion tops fall over (ripen) naturally; store under cold dry conditions; and do not allow cured bulbs to be exposed to sunscald, water, or high humidity.

Fusarium basal rot. Grow resistant varieties.

Pea

Seed rot and damping-off. Plant seed treated with captan or thiram.

Root rot: Plant early in well-drained soils. Use a fungicide seed treatment.

Fusarium wilt: Plant only varieties that are resistant to wilt.

Peppers

Seed rot and damping-off. Plant seed treated with captan or thiram.

Bacterial spot. Use the hot-water seed treatment, or dip seed in a solution of 1 part household bleach to 3 parts water and rinse with water. Follow with a fungicide seed treatment. Apply weekly sprays of a fixed copper fungicide such as COCS, through the bloom stage. Use crop rotation outside the tomato family

Tobacco mosaic. Plant resistant varieties.

Potato

Viruses, late blight, Verticillium wilt, and ring rot. Plant only "certified" disease-free seed.

Seedpiece rots. Treat cut seedpieces with captan as a dust or dip. Plant in warm soil (over 50°F) after the cut surfaces have corked over (suberized).

Early blight and late blight. Practice a 2- to 3-year crop rotation. Destory or remove cull tubers. Spray at weekly intervals, starting when the plants are 4 to 6 inches high with an approved fungicide. Plant late blight-resistant varieties. Verticillium wilt and root-knot nematodes. Plant resistant varieties.

Scab. Plant resistant varieties, or adjust the soil pH to 5.2 to 5.5. Use fungicide seedpiece treatments.

Storage rots. Store only disease-free, unbruised tubers. Allow tubers to cure at 60°F for 5 to 7 days before reducing temperature to 36° to 40°F (if possible).

Sweet corn

Seed rot and damping-off. Plant seed treated with captan or thiram.

Smut. Plant resistant varieties. Remove smut galls from the garden before they break. Do not compost smutted corn or infected corn residues.

Stewart's bacterial wilt. Plant resistant varieties. Control flea beetles with a recommended insecticide. Early season control is particularly important.

Foliage blights and rust. Plant varieties resistant to northern and southern corn leaf blights and rusts. Apply an approved fungicide. However, the use of fungicides is rarely justified. Diseased plant material should be removed from the garden. Blighted leaf tissue is safe to compost.

Maize dwarf mosaic. Plant as far from Johnsongrass as possible. Control wild Johnsongrass plants. Plant tolerant varieties.

Spinach

Seed rot and damping-off. Plant seed treated with captan or thiram.

Cucumber mosaic virus (blight). Plant resistant varieties.

Downy mildew or blue mold. Plant resistant varieties.

Sweet Potato

Black rot, scurf, foot rot, wilt, and soil rot. Buy certified, disease-free slips. Plant resistant varieties. Use a 3- to 4-year rotation.

Storage diseases. Do not expose roots to temperatures below 55°F. Cure the roots immediately after digging at 85°F and high humidity for 10 to 14 days. Store at 55°F. Never store in airtight bags or containers. Apply an approved fungicide as a post-harvest dip. Follow the manufacturer's directions carefully.

Tomato

Seed rot, damping-off, bacterial spot, bacterial canker, and tobacco mosaic virus. Plant seed which has been treated with hot water or acid and then treated with captan or thiram. Purchase seed or transplants only from reputable dealers. Soaking seed in copper sulfate solution (2 ounces per gallon) or dipping in bleach solution as described under peppers will help prevent bacterial spot and canker if seed are saved.

Bacterial spot: Plant certified, disease-free seed or transplants. Practice a 2-to 3-year crop rotation. Spray plants weekly with a fixed copper fungicide, beginning with the first sign of disease, through the bloom stage. Early blight, Septoria leaf blight, anthracnose, buckeye rot, and soil rots. Spray weekly with an approved fungicide. Practice 2- to 3-year crop rotation. Remove or destroy infected plant parts. Carefully follow label directions as regards the days between the last spray and harvest. Staking or caging plants and providing good air circulation will help reduce foliar diseases. Staking, caging or mulching will also help prevent fruit rots, such as anthracnose, buckeye rot, and soil rot.

Blossom-end rot: Maintain uniform soil moisture by mulching the plants and irrigating when needed. Avoid heavy pruning, excessive nitrogen applications, or damage to the roots by cultivation or hoeing.

Verticillium and Fusarium wilts: Grow only VF- or VFN-resistant varieties. There are many varieties available.

Root-knot nematodes: Grow VFN-resistant varieties.

Viruses: Avoid contact between potatoes and tomatoes to prevent "double infections" of potato and tomato viruses. Wash hands thoroughly with soap and hot water before handling your plants. Do not use tobacco when working with tomatoes. Most importantly, use proper cultural practices to promote and maintain healthy plants. Healthy plants do not get diseases as readily as weak ones. Healthy plants are the best control against plant diseases.

Natural Pest and Disease Control in Your Vegetable Garden 0

Controlling pests organically means much more than simply changing the types of sprays and dusts you use. Organic gardeners aim to create a balanced system where pests are regulated naturally and where there is seldom the need to use even the safest organic sprays and dusts.

One of the principles at the heart of organic gardening is to completely eliminate the need to use sprays and dusts, even organically acceptable ones. While this might seem foolish — or even impossible — at first, by removing the need to continually spray and dust your garden, producing your own food becomes more enjoyable, safer and will require less work. This means more time spent admiring and eating the fruits of your labour, and less time spent preparing concoctions and potions!

So where do you start on the road to organic, pest-free self-sufficiency? Surprisingly, some of the best ways to contol pests and diseases don't at first glance appear to be controls at all. By looking at her vegetable garden as a mini ecosystem which needs to be kept in balance, and thus in which no pests or disease can get out of control, the organic gardener is able to concentrate her efforts on:

- cultural techniques which will keep her vegetable plants in the best of health
- Natural Pest and Disease Control in your Vegetable Garden 0
- developing rich, biologically active soil
- Natural Pest and Disease Control in your Vegetable Garden 0
- encouraging a diverse population of predator species to keep insect pests in check.

But the place to start, and the foundation of your future organic vegetable growing success, starts with the plants themselves.

Buying Healthy Plants

Starting out with the healthiest plants you can find is a good way to ensure a healthy and productive vegetable garden. Use these tips to choose the best plants you can find:

Buy from nurseries or market stalls where the plants are well cared for. Plants which are allowed to stand in the sun on a hot pavement or which have been allowed to wilt are not a good choice, even if they're cheap.

Look at the entire selection of plants on offer. If some of the plants appear to be in poor health, it may be that the healthier looking ones have arrived more recently from the supplier and have yet to succumb to the seller's mistreatment. They might look healthy today, but could very well be diseased tomorrow.

Carefully look at the plant's colour. Plants which appear slightly pale might just need to be fertilized, and this is easily corrected. But plants which have distinct yellow streaks or brown spots are diseased. Do not buy them, or any other plants in the selection.

Look for roots, and look at the roots if possible. Plants which have masses of roots growing through their container's drainage holes have been left in too small a container for too long, and this could result in severe transplant shock if much of the root mass needs to be boken off in order to remove the plant from its container before transplanting. If you are able to (and if you are buying large quantities of plants, insist that you be allowed to) gently shake a plant loose from its container so that you can exmine the roots. The roots should be plentiful, but not wrapped in a tight spiral or almost filling the container.

Cultural Techiques

Water early: If you irrigate your garden with a sprinkler from overhead, it's best to water early in the day so plants can dry off before night falls. Foliage that stays wet for long periods of time is susceptible to leaf diseases, fungi that grow on leaves, tender stems, and flower buds. This tends to be a problem when plants stay wet throughout the night: Fungi spread quickly during the cool, moist evening hours. The fungi will cause the plant to be weakened, flowers will fall off, and fruit will begin to spot and become soft

Crop rotation: Do not grow the same plant family in the same spot year after year. Repetition of the same crop gives diseases a chance to build up strength. Design your garden so that each family of vegetables — cabbage family, cucumber family, and tomato/pepper family — can be moved to another block of your garden on a three-year rotation.

Avoid monocultures: Planting a wide variety of plants in your garden is another way to foil pests and diseases by encouraging diversity. Conventional lawns are an excellent example of a monoculture. Planting a mixture of lawn grasses will help prevent diseases from sweeping through your lawn, because some grasses will be susceptible and others won't.

Companion planting is another way to use diversity – and avoid monocultures – to foil pests. Mixing marigolds and herbs in and between your vegetable plants, as opposed to planting solid blocks or rows of a single vegetable crop, is very effective, because many pests locate crops by smell and these strong-smelling plants can "hide" a crop by masking the odour which attracts insects.

Developing Rich Soil

Soil that is fertile, well-drained and teeming with communities of a diverse range of microorganisms will greatly increase your vegetable garden's health and productivity, since many pests and diseases spend part or all of their lives below-ground. Having a diversity of microbes in the soil will help keep them in check.

Healthy soil is an intricate mix of tiny rock particles, organic matter, water, air, fungi, microorganisms and other animals. The more organic matter you add to your soil, the more biologically active your soil will become, and the more active it is, the greater the competition between benign microorganisms and those intent on doing harm to your plants. Over time, adding organic material improves the soil's structure, which in turn improves its ability to absorb and release water, thereby maintaining ideal moisture levels.

Although rich, well-fed soil is all most vegetable plants need to keep them growing vigorously and producing well, fertilizing them at the correct time is also beneficial, and will help boost their natural resistance to pests and disease. As a general rule, annual vegetables should be well-fed when they are young, and then given "booster" feeding when they start producing. Perrenial vegetables should be fertilised during the first third of their active growth in spring.

You can also fertilise them in summer, but not in autumn, as late fertilisation can stress them by encouraging tender new growth which may be winter-killed. When you apply fertiliser, always keep in mind that more is never better. Over-fertilisation leads to rank, spindly growth which always worsens problems with aphids and some other pests.

Encouraging Predators

Encouraging a diverse community of insect predators is one of the best ways to manage pests. Beneficials, the term used for insect predators and parasites, play a vital role in the complex community which naturally exists in your garden by eating or parasitizing harmful pests.

The first and most important rule for the gardener who wants to encourage beneficial insects is to immediately stop using toxic sprays and dusts. Even organic pesticides kill indiscriminately, so use them only when absolutely necessary and then only on the plants being attacked. To attract beneficial insects into your garden, provide them with an attractive food supply. Since the adults of many beneficial insects only feed on pollen and nectar, the best way to attract them is to plant small-flowered plants such as dill, fennel, parlsey and mint.

You can also allow a few of your vegetables, especially members of the mustard family such as radishes, and broccoli to go to flower. There are also many ornamental annuals which can be used: simply use them as colourful borders or edging plants around your vegetable beds. Once the adults of beneficial species have arrived in your garden and had a good meal, the females will search for good places to lay eggs which will hatch into predatory larvae. Beneficial ground-dwelling insects such as beetles find refuge in the soil, under stones and in thick mulches.

It is important to minimise dust and to provide a source of water in hot, dry areas to protect beneficial insects, which are easily killed by dehydration. An old birdbath filled with water and rocks (to provide safe landing places and prevent drownings) will be used by many beneficial insects.

Birds are amongst the most efficient predators of harmful insects, so planting shrubs and trees, and supplying them with a source of water to attract them to your garden – and encourage them to stay and eat their fill!- is a worthwhile investment.

And for consuming slugs, cutworms and many other ground-dwelling pests, nothing beats frogs and toads. Be sure to provide shelter and water for them, too.

Non-toxic Control of Insects

There are many different non-toxic control methods we can use against insects in the garden. Some are commercial sprays and powders, others can be made at home from household ingredients. Use non-toxic insecticides only when absolutely necessary The following list of non-toxic insecticides should only be used whco absolutely necessary. It's important to allow the garden to develop it's own micro-divcrsily and ccology.

Bacillus thuringiensis of the microorganisms that make insects ill, Bacillus thuringiensis is by far the best known. Though safe to humans, it has a devastating effect on many pests. BT, as it is often called, appears on the market in the form of a soluble powder. The powder is suspended in water and sprayed on plant surfaces, where it is ingested by the pest. The BT spores then germinate into plants that occupy more and more of the victim's body. A toxic crystal is also produced and is thought to be partially responsible for the insecticidal effect. BT is useful on both vegetable and fruit crops, aiding in the control of caterpillars, earworms, peach tree borer, cabbage moth, cabbage looper, and gypsy moth caterpillar.

Pyrethrum A is a botanical insecticide derived from the flowers of a species of chrysanthemum. The material causes rapid paralysis of most insects, but the insects usually recover unless the pyrethrum is combined with a synergist or other poison. There are several pyrethrmun products available that are certified organic. Use against aphids, thrips, caterpillars, ants, flies, earwigs, cabbage moth. As pyrethrum also affects beneficial insects it should only be used as a last resort.

Wettable sulphur is an effective fungicide and miticide. The sulphur interferes with important processes in fungal spores and is noted for its ability to suppress mite populations.

Chapter 5

Nematode Dynamics in Plants

Yield losses are influenced by the pathogenicity of the species of nematode involved, by the nematode population density at planting, by the susceptibility and tolerance of the host and by a range of environmental factors. Because of this, available models only estimate yield losses as proportions of the nematode-free yield. Estimating threshold levels further involves various economic calculations. Consequently, predicting yield losses and calculating economic thresholds for most nematode/crop problems is not yet possible. What is needed is more field-based information on the relationship between nematode population densities and crop performance, and various approaches to obtaining such data are described. Measuring the population density, especially of *Meloidogyne* species, is a major problem which needs addressing. Nematode population dynamics are also density dependent and are influenced by host growth, the reproductive potential of the species and by various environmental factors. Consequently, modelling nematode population dynamics is an equally impressive science. Again, good field data are required but the complicating effects of biological control agents, host susceptibility differences and environmental factors, and errors associated with measuring initial population densities, may mean it is practically impossible to predict reliably the multiplication rates of most nematodes, especially those with several generations per season.

Nematode Damage

The nematode, the host and the environment are the three interacting variables influencing the extent of yield loss in infested soils. An understanding of the mechanisms and principles involved in these interacting relationships is basic to being able to predict yield reductions from estimates of pre-planting nematode population densities (Pi).

Damage Models

When modelling the damage caused to plants by root-feeding nematodes certain basic principles apply. These are:

- Damage is proportional to the nematode population density.
- The degree of damage is influenced by environmental factors.
- The yield harvested is determined by the amount of light intercepted by the crop, by how efficiently the intercepted light is converted into dry matter, and finally by how that dry matter is partitioned into non-harvested and harvested yield. For some crops significant variations in moisture content will also affect final yield.

The above principles can be simply stated but are more complex in practice. Damage may be proportional to the nematode population density, but there are several qualifications of this statement. The relationship is usually curvilinear, increasing numbers of nematodes having proportionally diminishing effects. There is some evidence that at low densities the host plant can repair the damage and that growth may even be slightly stimulated. Seinhorst (1965) termed the population density (Pi) at which damage first became apparent as the tolerance limit (T).

Equally, at very high values of Pi, increasing numbers of nematodes may not further reduce dry matter productivity. Seinhorst termed this the minimum yield (m). There are various reasons why m may occur; there may be some growth before attack starts or after it finishes, and a significant biomass may be planted (e.g. potato tubers). However, m applies to total dry matter and because of effects on partitioning, the harvest value of m may be greater or less than that for total dry matter.

The third parameter in the Seinhorst equation is z, a constant slightly less than one. The equation is:

for Pi>T

y=1 where Pi£ T

where y is the yield.

An important qualification is that y is expressed as a proportion of the nematode-free yield. Hence, according to Seinhorst, the greater the yield potential the greater the loss in tonnes per hectare for any value of Pi.

The Seinhorst equation is usually plotted with Pi on a logarithmic scale, producing a sigmoidal curve. In practice T is usually small and the Pi value at which m is reached is so large that it is only the central part of the curve that is of practical use. Oostenbrink (1966) suggested that this approximated to a straight line. The equation for such a line is:

$$y = y(\text{max}) - \text{slope constant} \times \log \text{Pi}$$

Even the simplified Oostenbrink relationship is not very helpful. Yield is still expressed in proportional rather than real (tonnes per hectare) terms. Also, there is no way of applying the relationship without considerable experimentation to determine the slope of the regression.

The slope of the regression varies for several reasons. These include differences in pathogenicity (capacity to cause damage) between species, e.g. *Meloidogyne* spp. may be inherently more damaging than *Tylenchus* but we have no measure of their relative pathogenicities. Different plant species and varieties within species differ in their tolerance (capacity to withstand nematode damage). Also, there are large environmental influences on the damage suffered and particularly how that damage is translated into effects on final yield.

An important consideration, often overlooked, is the basis of measuring Pi. Usually it is given as numbers per gram of soil. A more appropriate measure is per unit volume of soil as this allows for bulk density differences. Numbers per gram of root is probably the most appropriate, but is difficult to measure because it is always changing. This latter aspect becomes important when trying to relate results from experiments where root densities are very different, e.g. pot and field trials.

A further problem is encountered when considering damage by nematodes that have two or more generations in the lifetime of a crop. Usually the Pi is measured at planting, but on a good host population of, for example, *Meloidogyne* spp., they can increase from below the value of T to a level in mid-season where they cause significant damage. Even so, it is a race between increasing Pi and increasing plant size that brings with it increasing tolerance (in Seinhorst terms, increasing m). In such situations suitability as a host (susceptibility) and tolerance can have a marked effect on the degree of damage.

In summary, both the Seinhorst and Oostenbrink equations are, without the addition of a substantial amount of additional information, purely descriptive and cannot be used to predict actual yield losses.

Mechanisms of Damage and Environmental Effects on Damage

Damage is proportional to the intensity of attack; this is often proportionally greater in sandy soils where nematodes can move more freely, than in heavier soils where movement is impeded. Adequate soil moisture is essential for free movement so attack is often limited as soils dry out later in the season. Temperature also influences the rate of nematode movement, but plant growth is usually equally affected.

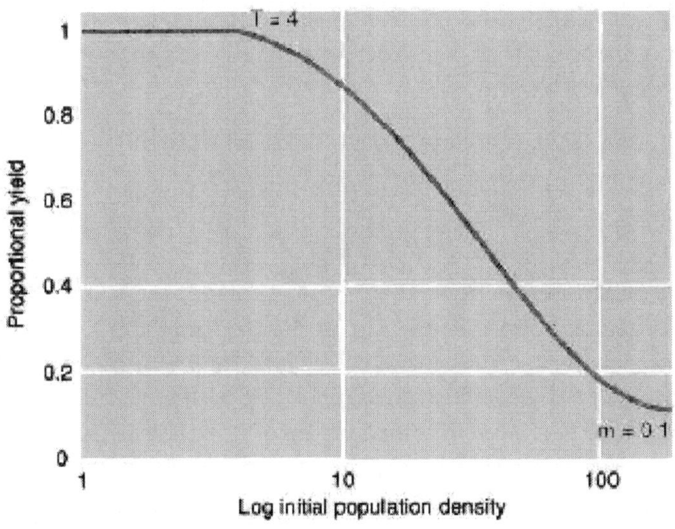

Figure: *The relationship between proportional yield loss and initial population density as modelled by Seinhorst (1965)*

Primary damage to the attacked roots can be attributed to mechanical damage associated with feeding or invasion, to withdrawal of nutrients, and/or to more subtle physiological effects. Generally damage reduces the rate of root extension. This reduces the rate of uptake of nutrients and water and, if any become limiting (and they usually do, even for crops without nematode damage), top growth rates are reduced. This reduces the rate of increase in light interception and carbohydrate synthesis and hence the capacity of the plant to generate more roots to overcome the limitations imposed by nematode damage. Such appears to be the main mechanism of damage by potato-cyst nematodes (*Globodera* spp.) whose effect is further increased by reductions in root efficiency, revealed in a decrease in root: shoot ratio. Further damage is associated with withdrawal of nutrients by the developing females (resistant cultivars of potato are often less damaged than susceptible cultivars) and by secondary pathogens such as *Verticillium dahliae*. The central role of nutrient uptake is revealed, however, by the substantial ameliorating effect on damage of additional fertilizer.

With *Meloidogyne* spp., impaired water relations appear to contribute substantially to reduced rates of top growth. This is probably because the developing giant cell systems interfere with and disrupt the developing xylem. Clearly, with such damage, effects on growth and yield are likely to be greater where the plants are on the threshold of becoming moisture stressed. Other effects include reduced photosynthetic efficiency and these are reviewed in Trudgill (1992).

Effects on Light Interception and Utilization

There is a good correlation in many crops between percent ground cover (i.e. the percentage of ground occupied by a plant or a crop, when viewed from above, that is covered by green leaves) and percent light interception. Most annual crops start as individual, separate plants and a reduction in growth rate is directly reflected in ground cover and hence light interception. As they grow the leaves of neighbouring plants merge to form a continuous canopy. Nematode damage that only delays the production of a continuous canopy, and hence 100 percent light interception, will have a smaller effect on final yield than damage which prevents the crop achieving such full cover. Premature crop death will also proportionally reduce yield.

Environmental Effects

Several environmental interactions have already been mentioned. Soil type clearly has an effect because it influences nematode movement as well as being a nutrient and water supply to the host. It can also influence nematode survival during periods of stress and will certainly influence the species composition of nematode communities. The effect of fertilizer practice and of water availability has also been mentioned, but these in turn will interact with host genotype and husbandry factors such as spacing and time of planting. Recent studies of potato-cyst nematodes illustrate some of the interactions and are briefly summarized below:

- The interaction between two potato cultivars of different tolerance and rates of compound fertilizer and the nematicide aldicarb was studied at a site with a sandy soil. In this trial, the site was uniformly heavily infested with *Globodera pallida* and the tolerant cv. Cara produced tops that were generally twice the size of intolerant cv. Pentland Dell. Consequently, Cara tended to produce many more leaves than were required to give 100 percent ground cover. The yield of the Cara was increased equally by a half and a full rate of aldicarb whereas that of the Pentland Dell was increased more by the full rate. Similarly, increasing rates of fertilizer proportionally increased the yield of Pentland Dell untreated with nematicide more than it did that of treated Pentland Dell or untreated Cara. This trial and several others showed that initially the *G. pallida* proportionally decreased the top growth of both cultivars to the same degree, supporting the basic proportional model proposed by Seinhorst.
- A series of five trials on different soil types tested the same five potato genotypes in plots with a wide range of initial populations (Pi) of *G. pallida*. Excellent regressions between Pi and tuber yields were produced revealing differences in tolerance between genotypes and in overall rates of yield reduction at the different sites. Further analysis showed that variations from a basic model similar to a simplified Seinhorst curve (without T or m) could be partitioned into genotype and

site effects. The former were common across sites and the latter across genotypes.

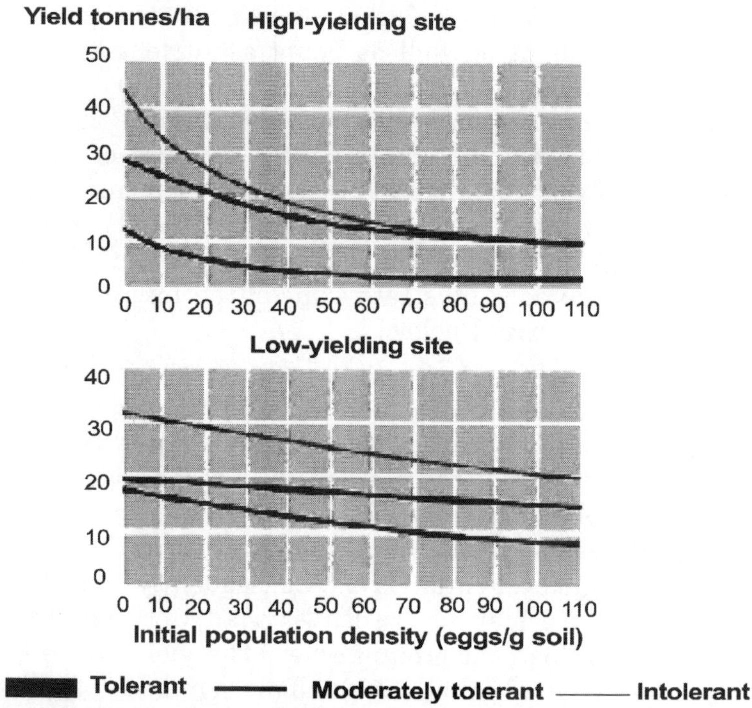

Figure: *The relationship between initial population density of* Globodera pallida *and tuber yield for tolerant, moderately tolerant and intolerant genotypes at two sites with contrasting yield potential*

This information provides the basis for predicting the effects of *G. pallida* on the tuber yields of different cultivars classified on their degree of tolerance and of sites classified by their soil type. However, the losses are still predicted as a proportion of the nematode-free yield. To have a prediction of the actual loss in tonnes per hectare requires an estimate of the yield potential of the cultivar and site, which requires yet further modelling.

Only with this information can yield losses be accurately quantified in financial terms and the tolerance limit identified. The alternative is to extrapolate from the available trial data and make allowances on the basis of experience for the obvious possible environmental influences. Methods of estimating yield loss are therefore of central importance and are considered below.

Methods of Estimating Yield Losses

Pot studies can be used to determine some of the basic information on yield-loss relationships, but because of environmental differences and interactions, field studies are also needed. There are two approaches: one is to use nematicides at relatively uniformly infested sites; the other is to work at sites with a range of population densities but which are uniform in other respects. A combination of both approaches is often a happy compromise.

The former gives practical information on the effectiveness and potential value of a particular treatment but tells little about the nature of the relationship. It also suffers from the criticism that nematicides have a range of side-effects. The latter has the benefit of producing information on the relationship between Pi and yield, but it requires experimental errors to be minimized. Because Pi estimates have large errors, accuracy is improved by reducing plot size and by taking and processing multiple samples from each plot. However, plot size must be large enough to obtain a realistic yield and adequate guard plants are essential.

Another option is to establish many small plots in large but otherwise uniform fields. These can be at random, in a grid pattern or along known trends in Pi. The plots can be split and a nematicide applied to one half. For each plot the Pi and yield are determined. The results will produce a scatter of points, hopefully with yield decreasing as Pi increases. Much of the scatter is due to errors in estimating Pi and yield, and it can be minimized by taking the average of all the results within each error band. Such an approach needs:

- a wide range of initial populations;
- a uniform field;
- a large number of plots (100 or more); and
- the plots to be part of an otherwise uniform crop.

Control Measures

Control measures aim to protect the treated crop from damage, and to prevent nematode multiplication and so reduce the threat to the next susceptible crop in the rotation. Most cost-effective and successful is the growing of resistant varieties.

However, while these will prevent nematode multiplication, they are often as vulnerable to damage as a susceptible variety. In yield-loss studies resistant varieties can be a very useful tool for preparing plots with reduced populations without the side-effects associated with other treatments. Between vulnerable crops, rotations involving non-hosts are almost essential. Nematicides, whether natural or artificial, are a last resort and should not be used as a crutch to compensate for poor management. They are always costly and frequently toxic and environmentally damaging. However, their side-effects can make them attractive in some situations; the oximecarbamates control a broad range of pests, until they develop resistance, while the fumigant nematicides release nitrogen, further increasing yields.

Population Dynamics

Nematodes have various reproductive strategies. Some grow large and have long life cycles with low rates of population increase (K strategists), others are relatively small, have short life cycles and potentially higher reproductive rates (r strategists).

An endoparasitic habit with induction of giant cells or other rich and continuously available food sources, reduces exposure to predation and other stresses and further increases reproductive potential. A reduction in the number of active juvenile stages further decreases development time, thereby reducing generation time and increasing the potential for multiple generations in a season. A wide host range completes the adaptation of pathogens such as some *Meloidogyne* spp., which can be regarded as the ultimate plant-parasitic nematode r strategists. Many *Longidorus* spp. are examples of K strategists.

It is a characteristic of K strategists that they do best in stable environments where populations are usually close to the equilibrium density (the population density that can be sustained). In contrast r strategists increase rapidly where the environment is favourable, often overshooting the equilibrium density. Severe damage to the host occurs and the population crashes. This can occur with repeated cropping of hosts as Pi increases, environmental influences on sex determination reduce multiplication, parasites of the nematode increase in number, and increasing damage to the host and competition for feeding sites progressively reduces multiplication.

Consequently, nematode multiplication rates are strongly density dependent. Again, the question of how density is defined arises. Usually it is expressed as the numbers of nematodes per gram or ml of soil, but the units that directly affect the nematode are those that are root related, e.g. number of root tips and/or length or weight. Hence, a cultivar with twice the root mass of another will, except at low densities where the multiplication rate is the maximum, support a higher multiplication rate. Similarly, tolerant cultivars that maintain a greater root mass as Pi increases than intolerant cultivars, will have a greater equilibrium density and maintain a greater multiplication rate at high pre-planting population densities.

Overall multiplication rates are determined by the intrinsic maximum rate of multiplication, which is influenced by nematode species, the susceptibility (defined as all those qualities favouring the nematode) of the host, and the various environmental factors that influence both the nematode and the host. Nematode multiplication can be modelled in different ways. For migratory nematode species that multiply continuously, Seinhorst (1966) proposed the following formula derived from a logistic equation:

$$Pf = aEPi/(a - 1)Pi + E$$

where a is the maximum rate of increase and E is the equilibrium density at which Pf = Pi. For sedentary nematodes with one generation at a time, e.g. potato-cyst nematodes, Seinhorst (1967) proposed an alternative model based on the competition model of Nicholson (1933):

$$Pf = a\left(a - q^{Pi}\right) / -\ln q$$

where a is again the maximum rate of multiplication, and 1-q is the proportion of the available space which is exploited for food at a density of Pi= 1.

Jones and Perry (1978) also proposed a model for sedentary nematodes with a logistic basis derived from the observation that sex determination is density dependent. Their model includes parameters that reflect fecundity and the proportion of the population that does not hatch.

All three models, in their most basic form, show maximum rates of multiplication at low initial densities. As Pi increases the rate of multiplication is reduced as an upper asymptote is reached. In reality, the shape of this curve is modified as Pi increases due to the increasing damage inflicted and the loss of roots. With the Jones and Perry model this is exacerbated as space lost as a result of root damage increases the competition between invading nematodes, resulting in an even greater shift in the sex ratio towards male production than would otherwise be the case. Thus the approach to the asymptote is slower and indeed the asymptote is reduced below the theoretical level. Further increases in Pi can inflict so much root damage that the population increase becomes negative and the population size is ultimately reduced. All the equations mentioned require modifying by including a damage function such as that of Seinhorst, which also allows the differences in tolerance between cultivars to be taken into account. The damage functions used model proportional differences, and further modification may be required to account for absolute differences in plant size.

Another plant characteristic that affects population increase is the host status of the plant. Differences can be modelled in terms of the maximum multiplication rate or the space required for successful multiplication (Seinhorst models), or in terms of fecundity or effects on the sex ratio.

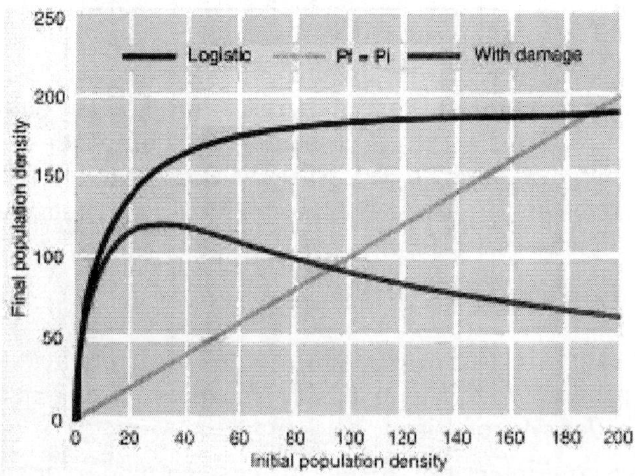

Figure: *The theoretical logistic relationship between initial population density and final population density and the relationship when roots are damaged*

An important method of expressing and comparing the effects of different cultivars or cropping regimes is to consider the equilibrium density, i.e. the point at which Pf = Pi. This density is usually observed at a Pi which is larger than that which gives the largest Pf. In practice this equilibrium density is reached after a period of oscillation about the equilibrium density. The size of the oscillations will be determined by the tolerance and resistance of the host. Tolerance and resistance will produce small oscillations, while susceptibility and intolerance can result in large oscillations. Indeed, these two factors can interact to the extent that a tolerant but partially resistant cultivar can produce a higher equilibrium density than an intolerant susceptible cultivar.

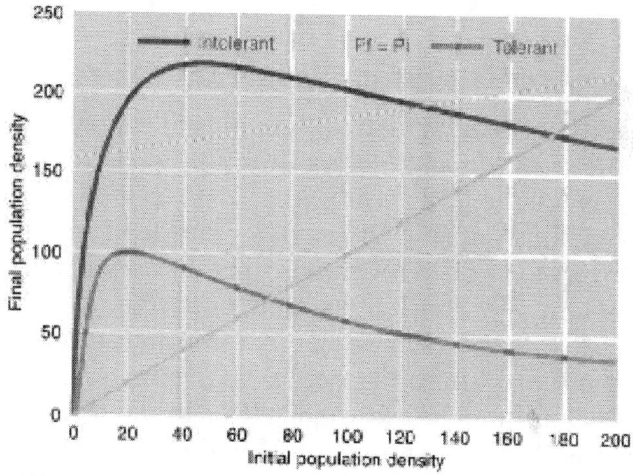

Figure: *The relationship between Pf and Pi when a tolerant and an intolerant host are grown*

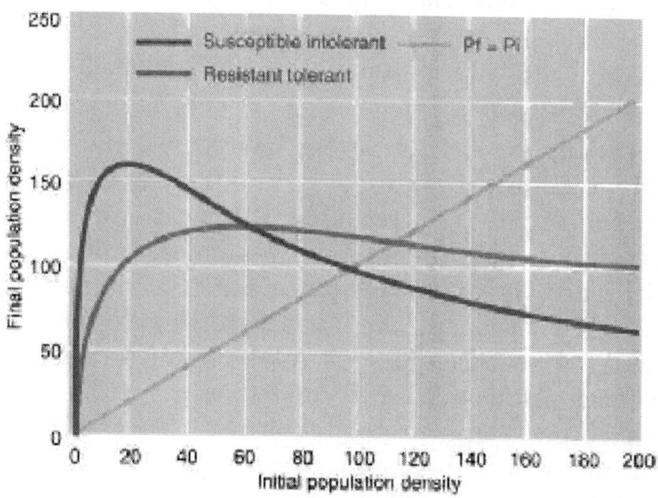

Figure: *The relationship between Pf and Pi contrasting the response when an intolerant susceptible host is grown to that which occurs when a tolerant but partially resistant host is grown*

Care needs to be taken in devising management strategies for the control of nematodes to balance the benefits of tolerance against the benefits of resistance, to ensure that while yields are maximized, nematode populations are not raised to levels that are damaging to other cultivars. Models can be used to examine and explore nematode management strategies, but need to take into account the effective population if this is less than the actual population, and the decline in the numbers of nematodes in the absence of a host crop.

Nematology and Nematode Introductions

Present-day agricultural development relies on the generation of new technologies through research and their adoption by farmers. In the Near East region agricultural scientists are attempting to provide more food in sustainable systems to meet the needs of growing populations. This situation poses tremendous challenges and opportunities for those involved in plant protection research. Plant-parasitic nematodes are a severe constraint on agricultural production. They cause considerable annual losses among the broad range of crops grown in the Near East region. Major inputs are needed to reduce the damage caused by these pests, and increase the world's food supplies. Progress in agricultural nematology research in the Near East region has been notable in the last 30 years. Nematology, as a new science within the framework of plant protection disciplines, was established in the region around the mid-1950s.

Since then gradual progress has been achieved through extensive teaching programmes and research projects. At present, the number of nematologists in the region, whether they are affiliated to universities, academies, and/or ministries of agriculture, is quite adequate for the advancement of nematology education, research and technology in the respective countries. Few research projects related to nematode problems have been sponsored by foreign donors for promoting knowledge about these pests and protecting crops from their damaging effects. Information on nematode diagnosis, biology, population dynamics, and host-parasite relationships has been acquired as a valuable prerequisite for the future establishment of management control practices. The fact that countries in this region have different ecological zones with various cropping patterns that influence the nature and magnitude of nematode problems emphasizes the need to strengthen their national research and extension capacities.

Nematode Problems Encountered in the Region

Nematode problems are aggravated by prevailing environmental conditions that favour rapid reproduction of certain species. Infestation appears to be widespread where cultivated lands in the region are concentrated in specific ecological zones:

- In temperate zones at intermediate to high altitude, with high rainfall, snow and cold winters, and dry cool summers, the cyst nematodes, *Globodera rostochiensis, Heterodera schachtii, H. trifolii* and *H. avenae* infecting potato, sugar beet, clover and wheat, respectively, are quite prevalent.
- In moderately warm zones at intermediate altitude, with cool rainy winters and hot dry summers, especially in the coastal belts, the root-knot nematodes, *Meloidogyne incognita, M. javanica, M. arenaria, M. hapla, M. thamesi, M. naasi* and *M. artiellia* commonly occur in many field and horticultural crops; certain cyst-forming nematodes e.g. *Heterodera latipons* on wheat and barley, *H. ciceri* on chickpeas, *H. goettingiana* on broad beans and peas, *H. cajani* on cowpeas, and *H. zeae* on maize, are frequently found; the stem nematode race of *Ditylenchus dipsaci* attacks broad beans and various species of the root-lesion nematodes, especially *Pratylenchus penetrans, P. thornei, P. neglectus* and *P. vulnus,* are widespread in many orchard, field and vegetable crops; and the citrus nematode, *Tylenchulus semipenetrans* is a common pest in citrus, olive and vineyard plantations.
- In warmer and subtropical zones at low altitude, with cool winters, little rainfall and very hot humid summers, the root-knot nematodes, *M. javanica, M. incognita* and the root-lesion nematode, *P. brachyurus* are frequently found in date-palms, olives, mangoes, groundnuts, alfalfa, melons, tomatoes, okra and beans; the root-rot

nematodes, *Radopholus similis* and *Helicotylenchus multicinctus* are commonly associated with the decline of banana and sugarcane plantations; the reniform nematode, *Rotylenchulus reniformis* is widespread in the barbadense cotton belts, banana plantations and many vegetable-growing areas; and the rice nematode, *Hirschmanniella oryzae* is common in the paddy rice-growing belts.

These comments only summarize the extent of the distribution of the known economically important nematodes in the region. There are other nematode species that have been reported in countries of the region that have specific hosts as well as wide host ranges. Species of the genera *Longidorus, Xiphinema* and *Trichodorus* that are capable of transmitting viruses from one plant to another are also common.

Although many of the nematodes identified are capable of inflicting direct severe damage to growing plants, the degree of injury depends on the level of soil infestation by the dominating pathogenic species, new races, or pathotypes, and the prevailing environmental stress conditions such as drought, waterlogging, salinity, and inadequate plant nutrition. Many nematodes are involved in the expression of disease complexes, syndrome problems and in the reduction of benefits from nitrogen-fixing bacteria.

Impact on Productivity and Plant Protection Resources

Many plant-parasitic nematodes are widely distributed in this part of the world. Their association with other soil pathogens in disease complexes, as well as with abiotic stress conditions in crop physiological syndromes, ranks them in the list of major economic pests affecting the yield and quality of our major food and fibre crops. Unfortunately, the majority of growers in this region, being unaware of the nature and harmfulness of nematode infestations in their fields, do not seriously consider the destructive effect of these noxious pests on their crops. These pests are a severe constraint to agricultural production in several countries of this region. There are no reliable estimates of the magnitude of the annual losses to crops, but diagnostic surveys have shown, however, that many economic nematode species are widespread and the damage they cause to crops is more evident than was anticipated. Nematode damage in this region may well come up to the estimate of 10 percent of total production, and if these pests are not controlled, losses in the production of principal crops may be more serious.

Currently, many agricultural chemicals used for nematode control are no longer available because of health and environmental hazards associated with their use. Those nematicides that remain on the market are less effective and continue to increase in cost while farm revenues decrease. These realities demand that nematode management should become an integrated programme of practices that includes alternative measures to the use of chemicals and provides sustainable nematode control systems. There is no standard formula for a plant protection programme, and such programmes must be designed to fit the requirements and options available within a specific growing region.

Each farming situation requires a different set of production and protection management inputs for greatest profitability, and the requirements may change from year to year. Climate, soil type, cropping system, nematode pest history, cultural practice, cultivar and the nature of the surrounding environment all affect nematode problems on a given site. In some areas, where resistant cultivars have been deployed, their effectiveness has been rapidly reduced by the emergence of new host races of the target nematode species, or by emergence of new infestations of different nematode species on the resistant hosts.

These are but a few of the biological and environmental factors that illustrate the complexities of the nature of these pests in the region. It is a situation that calls for appropriate nematode management approaches. Modified cultural practices, non-host rotation crops, host-crop resistance and/or tolerance, biorational nematicidal compounds, nematode antagonists, cover crops and better nematode identification and enumeration are all tools that call for improvement as components of nematode management programmes. Various plant protection resources are available in this region and their potential is beginning to be explored. The following are examples of existing resources:

- *Human Resources:* The region has a number of highly qualified, university graduate specialists in plant protection and pesticide management disciplines. Those involved in nematology have acquired the necessary basic and applied studies in nematode taxonomy and biology that allow them to perform diagnostic surveys and research on nematode control strategies.

- *Energy Resources:* The region is particularly rich in fossil deposits that are exploited as petroleum and coal. New classes of chemical materials and by-products have been introduced and utilized as soil conditioners and mulches for crop production and protection.
- *Natural Botanical Resources:* Many areas in the region are endowed with native lines of wild plants that could be utilized as sources of genetic material in breeding for nematode resistance or as sources of biologically active chemicals that may be inhibitors or have nematostatic or plant-growth regulatory properties.
- *Natural Microbial Resources:* Sites with soil suppressive to certain nematodes have been identified. Isolates of bacteria and fungi antagonistic to nematodes have been discovered. Technologies to stimulate the effectiveness of such indigenous nematode antagonists are being explored through the employment of various soil amendments and organic additives or supplements.

Future Opportunities

In a meeting like this, we should be concerned about the future effectiveness of research programmes in reducing losses caused by nematodes and in developing and advocating appropriate agricultural systems that are sustainable, environmentally friendly and more productive within existing resources. Donor agencies should place strong emphasis on increasing nematology research capacity in this region, both through funding of projects by institutional grants and support of advanced training programmes. Institutional mechanisms, for example, such as those already established in some countries of the region as national nematology research centres, need to be improved and strengthened to act as centres of excellence for the advancement of agricultural nematology in the region. These centres of excellence should be adequately supported financially and technically to:

- collaborate with national programmes and national agencies in the design of projects related to basic mechanisms of nematode interactions in the agro-ecosystem, as well as biochemical relationships that may be of an allelopathic nature;
- facilitate research development and implementation of national nematode pest management programmes through technical support and distribution of materials for nematological research;
- engage in training and delivery of information in order to ensure more efficient use of resources;
- develop databases that quantify the relationship between nematode populations and crop-yield losses under the prevailing biotic and abiotic factors.

There is no doubt that there are many critical issues, specific projects and opportunities for which, during our discussions and consultations, we can develop adequate programmes with the aim of promoting nematological research and technology in this region of the world.

Chapter 6

Nematodes Management

Nematodes are tiny, wormlike, multicellular animals adapted to living in water. The number of nematode species is estimated at half a million, many of which are "free-living" types found in the oceans, in freshwater habitats, and in soils. Plant-parasitic species form a smaller group. Nematodes are common in soils all over the world. If all the matter in the universe except the nematodes were swept away, our world would still be dimly recognizable, and if, as disembodied spirits, we could investigate it, we should find its mountains, hills, vales, rivers, lakes and oceans represented by a thin film of nematodes.

An important part of the soil fauna, nematodes live in the maze of interconnected channels—called pores—that are formed by soil processes. They move in the films of water that cling to soil particles. Many genera and species have particular soil and climatic requirements. For example, certain species do best in sandy soils, while others favour clay soils. Nematode populations are generally denser and more prevalent in the world's warmer regions, where longer growing seasons extend feeding periods and increase reproductive rates. In the southern United States, as many as ten generations are produced in one season. Light, sandy soils generally harbour larger populations of plant-parasitic nematodes than clay soils. This is attributable to the more efficient aeration of sandy soil, the presence of fewer organisms that compete with and prey on nematodes, and the ease with which nematodes can move through the root zone. Also, plants growing in readily drained soils are more likely to suffer from intermittent drought, and are thus more vulnerable to damage by parasitic nematodes. Desert valleys and tropical sandy soils are particularly challenged by nematode overpopulation.

Plant-parasitic nematodes—the majority of which are root feeders, completing their lifecycles in the root zone—are found in association with most plants. Some are endoparasitic, living and feeding within the tissue of roots, tubers, buds, seeds, etc. Others are ectoparasitic, feeding externally through plant walls. A single endoparasitic nematode can kill a plant or reduce its productivity, while several hundred ectoparasitic nematodes might feed on a plant without seriously affecting production. A few species are highly host-specific, such as *Heterodera glycines* on soybeans and *Globodera rostochiensis* on potatoes. But in general, nematodes have a wide host range.

Endoparasitic root feeders include such economically important pests as the root-knot nematodes (*Meloidogyne species*), the cyst nematodes (*Heterodera species*), and the root-lesion nematodes (*Pratylenchus species*). Important ectoparasitic root feeders include: root, dagger (*Xiphinema*), needle (*Longidorus*, *Paralongidorus*), ring (*Criconemella*, *Macroposthhonia*), stunt (*Tylenchorhynchus* and *Merlinius*), pin (*Paratylenchus*), and spiral (*Helicotylenchus, Rotylenchus,* and *Scutellonema*) nematodes. Direct feeding by nematodes can drastically decrease a plant's uptake of nutrients and water.

Nematodes have the greatest impact on crop productivity when they attack the roots of seedlings immediately after seed germination. Nematode feeding also creates open wounds that provide entry to a wide variety of plant-pathogenic fungi and bacteria. These microbial infections are often more economically damaging than the direct effects of nematode feeding.

Major Plant-parasitic Nematode Genera in the U.S. and Associated Damage to Plants

- Root-knot nematodes (*Meloidogyne species*) form galls on injured plant tissue. The galls block water and nutrient flow to the plant, stunting growth, impairing fruit production, and causing foliage to yellow and wilt. Roots become rough and pimpled and susceptible to cracking.
- Cyst nematodes (*Heterodera species*) give plants an unthrifty or malnourished appearance, and cause them

to produce smaller-than-normal tops. Foliage is liable to wilt and curl, while roots become thick and tough and take on a red or brown colouring.

- Sting nematodes (*Belonolaimus species*) are found mainly in the South, especially in sandy soils with meager organic-matter content. Areas of stunted plants are an early indicator. As these areas grow larger and finally meet, the plants that were first affected will start to die at the margins of older leaves.
- Root-lesion or meadow nematodes (*Pratylenchus species*) cause internal browning in potato tubers and in the roots of corn, lettuce, peas, carrots, tomatoes, and brassicas.

Nematode control is essentially *prevention*, because once a plant is parasitized it is impossible to kill the nematode without also destroying the host. The most sustainable approach to nematode control will integrate several tools and strategies, including cover crops, crop rotation, soil solarization, least-toxic pesticides, and plant varieties resistant to nematode damage. These methods work best in the context of a healthy soil environment with sufficient organic matter to support diverse populations of microorganisms. A balanced soil ecosystem will support a wide variety of biological control organisms that will help keep nematode pest populations in check.

Symptoms and Sampling

Usually, sampling is done because the grower observes a section of field with unhealthy plants, or notices an unexplained yield reduction. Because nematodes damage roots, any condition that stresses the plant—such as drought (or even hot spells), flooding, nutrient deficiencies, or soil compaction—will tend to amplify the damage symptoms noted above. Failure to respond normally to fertilizers and slower-than-normal recovery from wilting are signs of nematode infestation. In the undisturbed soil of groves, turf, and pastures, visible symptoms of nematode injury normally appear as round, oval, or irregular areas that gradually increase in size year by year. In cultivated land, nematode-injured spots are often elongated in the direction of cultivation, because nematodes are moved by machinery.

It is important to note that species of nematode are present in all soils; their mere presence does not necessarily mean that they are damaging plants. Harmless or even beneficial species are found in proximity to plants, right along with the parasitic species. Beneficial nematodes feed on such pests as Japanese beetle grubs and plant-parasitic nematodes, and release nutrients into the soil by eating bacteria and fungi. An experienced nematologist can identify species, and determine which, if any, are responsible for the observed damage.

Nematode sampling techniques vary depending on the crop, the root depth, the type of nematode causing damage, and the time of the season. The procedure presented here is a generic sampling technique for annual crops. Soil samples taken in the late summer are best when testing for the presence of nematodes. Root-zone soil samples are best taken immediately after harvest, or just prior to harvest if the crop showed signs of damage. First, fields should be divided into 20-acre blocks that have similar damage, soil texture, or cropping history. From each block take several sub-samples, mixing them well to create a single one-quart sample for each block. Soil samples should be kept cool, but not frozen.

Samples for established perennial crops are best taken from the feeder root zone, which is usually located around the canopy drip line. Your county or state Cooperative Extension Service can provide names of commercial labs that have nematode-identification services.

Preventing Further Spread of Nematodes

Preventing nematodes from entering uninfested areas is important; under their own steam they can spread across a field at a rate of three feet per year. The following measures will help prevent human-assisted spread of nematodes to uninfested fields:

- Use certified planting material
- Use soilless growing media in greenhouses
- Clean soil from equipment before moving between fields (washing equipment—including tires—with water is most effective)
- Keep excess irrigation water in a holding pond so that any nematodes present can settle out, pump water

from near the surface of the pond; plan irrigation to minimize excess water

- Prevent or reduce animal movement from infested to uninfested fields
- Compost manure to kill any nematodes that might be present, before applying it to fields
- Eliminate important weed hosts such as crabgrass, ragweed, and cocklebur.

Managing Soil Biology

The basis of sustainable nematode control is the maintenance of a healthy soil food-web. This begins with routine application of organic matter. There is substantial evidence that the addition of organic matter in the form of compost or manure will decrease nematode pest populations and associated damage to crops. This could be a result of improved soil structure and fertility, alteration of the level of plant resistance, release of nemato-toxins, or increased populations of fungal and bacterial parasites and other nematode-antagonistic agents. Reduced nematode damage from increased organic matter in soil is likely a combination of these interaction. Higher organic matter content increases soil's water-holding capacity, and supports thriving communities of the decomposers and predators that make up the soil's "digestive system."

Nematodes are important participants in this underground energy-transfer system. They consume living plant material, fungi, bacteria, mites, insects, and each other, and are themselves consumed in turn. Some fungi, for example, capture nematodes with traps, sticky knobs, and other specialized structures. Nematodes and protozoa regulate mineralization processes.

Evidence suggests that between 30 and 50 percent of the nitrogen present in crop plants was made available by the activity of bacteria-consuming nematodes. Research in Denmark has indicated that nematodes convert about as much energy as earthworms in certain forest soils. Don't forget, the vast majority of nematodes found in the soil are not plant parasites.

The food-web's stability is challenged by the yearly turning of the soil, which reduces the numbers of organisms that displace or prey on plant-parasitic nematodes, while bringing more nematodes to the surface from deeper soil. If the same host crop is planted year after year, plant-parasitic nematodes may increase to damaging levels. Root-feeding nematodes are very opportunistic, and are among the first organisms to invade after a disturbance.

Keeping these facts in mind, it is important to actively manage soil biology using minimum-tillage practices, compost, animal manures, green manures, cover crops, and crop rotations. These practices help promote the growth of beneficial organisms while suppressing plant parasites. Certain organisms that are associated with well-managed crop soils—e.g., Rhizobacteria and mycorrhizae—may induce systemic host resistance to nematodes and to some foliar diseases.

Soil Amendments for Nematode Control

Some sources of organic matter known to be nematode-suppressive include oilcakes, sawdust, sugarcane bagasse, bone meal, horn meal, compost, and certain green manures.

Most nematode species can be significantly reduced by tilling in chitinous materials such as crushed shells of crustaceans. This is effective because several species of fungi that "feed" on chitin also attack chitin-containing nematode eggs and nematodes. Increasing the amount of chitin in the soil will also increase the population of these fungi. A shrimp-shell-based fertilizer called Eco Poly 21™ Micro shrimp fertilizer is available from Peaceful Valley Farm Supply. At 2002 catalogue prices, it would cost between $87 and $216 to treat an acre with this product (the suggested application rate is 20 to 50 lbs. per acre). Clandosan™, a nematicide made of crab shells and agricultural-grade urea, can be used as a pre-plant treatment.

Crop Rotations and Cover Crops

Crop rotation to a non-host crop is often adequate by itself to prevent nematode populations from reaching economically damaging levels. However, it is necessary to positively identify the species of nematode in order to know what plants are its host(s) and non-hosts. A general rule of thumb is to rotate to crops that are not related to each other. For example, pumpkin and cucumbers are closely related and rotating between them would probably not be effective to keep nematode populations down. A pumpkin/bell pepper rotation might be more effective. Even better is a rotation from a broadleaf to a grass. Asparagus, corn, onions, garlic, small grains, Cahaba white vetch, and Nova vetch are good rotation crops for reducing root-knot nematode populations. Crotalaria, velvet bean, and grasses like rye are usually resistant to root-knot nematodes. Rotations like these will not only help prevent nematode populations from reaching economic levels, they will also help control plant diseases and insect pests.

Allelochemicals are plant-produced compounds (other than food compounds) that affect the behaviour of other organisms in the plant's environment. For example, sudangrass (and sorghum) contain a chemical, dhurrin, that degrades into hydrogen cyanide, which is a powerful nematicide. Some cover crops have exhibited nematode-suppressive characteristics equivalent to aldicarb, a synthetic chemical pesticide.

Farmers in Alabama have added sesame into rotation with cotton, peanuts, and soybeans. Nematode levels are reduced and yields significantly increased among those crops in fields previously planted in sesame. Sesame yields averaged 1500 lbs per acre, well above the world average of 500 to 600 lbs per acre. Research shows that sesame may be an effective rotation crop to control peanut root knot nematode (*Meloidogyne arenaria*) and southern root knot nematode (*Meloidogyne incognita*). Sesame rotation is not effective, however, for the Javanese root knot nematode (*Meloidogyne javanica*). Commercial nematode control products derived from sesame include Dragonfire™ (oil), Ontrol™ (seed meal)—both manufactured by Poulenger USA—and Nemastop™ (ground up sesame plant) from Natural Organic Products.

In South Texas, soybean varieties were shown as possible alternatives to grain sorghum in cotton cropping sequences. Eighteen soybean varieties of maturity group 5, 6, 7, and 8 were tested in *Rotylenchulus reniformis*-infested soil, either nonfumigated or fumigated with 1,3-dichloropropene. Reproductive rates of R. *reniformis* were compared in the first year. Both experiments were planted with cotton in the second year to measure the rotational effects of soybean on cotton yield compared with grain sorghum and fallow. The high-yielding soybean cultivars with potential to suppress reniform nematode were "HY574," "Padre," "DP7375RR," and "NK83-30."

A 2000-2002 Maryland study evaluated crop rotations and other cultural practices to manage southern root-knot nematodes and lesion nematodes. Researchers grew nematode-susceptible potatoes and cucumbers, and compared the effect of several summer rotations on nematode problems. A summer rotation of sorghum sudangrass (*Sorghum bicolor x Sorghum arundinaceum var. sudanense*) reduced the root knot nematode population as effectively as the control treatment (soyabean cultivar with no known root-knot resistance and one nematicide application). Poultry litter/tillage (Year 1) and fallow (Year 2) were equally effective in managing the lesion nematode population. To maintain the effect, the rotations had to be included annually. Either summer or early-autumn sampling dates were more effective than midspring to identify threshold levels of the pests.

Nematodes and pH

Cyst nematodes do not hatch well in very acid soils (pH 4) or alkaline soils (pH 8). They do best in soil with a near-neutral pH of 6. This can be used to some advantage. For example, potatoes may be safest from nematode damage in an acid soil, while cabbage and beets can be planted in alkaline soil. But most plants do best at the pH that favours nematodes.

Researchers have observed that brassicas have a nematode-supressive effect that benefits the following crop in a rotation. This "mustard effect" is attributed to glucosinolate compounds contained in brassica residues. Toxicity is attributed to enzymatically induced breakdown products of glucosinolates, a large class of compounds known as isothiocyanates and nitriles that suppress nematodes by interfering with their reproductive cycle. These glucosinolate breakdown products are similar to the chemical fumigant VAPAM® (metam sodium), which degrades in soil to methyl isothiocyanate. Glucosinolate compounds are also responsible for the pungent flavours and odours of mustards and horseradish. Jack Brown, PhD, a plant breeder specializing in brassicas at the University of Idaho, has released two biofumigant varieties, "Humus" rapeseed and "IdaGold" mustard, each containing elevated levels of glucosinolates. Cover crop seed for mustards, rapeseed, and oilseed radish are available from a variety of sources. Several Extension Service bulletins describe the use of brassica cover crops in greater detail.

Allelopathic Cover Crops

Some plants produce allelochemicals that function as nematode-antagonistic compounds, such as polythienyls, glucosinolates, cyanogenic glycosides, alkaloids, lipids, terpenoids, steroids, triterpenoids, and phenolics, among compounds from these plants—e.g., castor bean, chrysanthemum, partridge pea, velvetbean, sesame, jackbean, crotalaria, sorghum-sudan, indigo, tephrosia—are exuded during the growing season or released during green manure decomposition. Sunn hemp, a tropical legume, and sorghum-sudan, a prolific grass plant grown for its biomass, are popular nematode-suppressive cover crops that produce the allelochemicals known as monocrotaline and dhurrin, respectively.

Here are some examples of how brassica crops are being used to manage nematodes:

- Oil radish as a green manure has dramatically reduced stubby root nematode (*Trichodorus*) and root lesion nematode
- Oil radish is used as a "trap crop" for the sugarbeet cyst nematode, its roots exude chemicals that stimulate hatching of nematode eggs. The larvae that emerge are unable to develop into reproductive females—reducing the population densities for the following crop.
- Rape or mustard plantings in rotation with strawberries have checked the increase of some nematodes.
- Rapeseed and sudangrass green manures grown prior to potatoes at Prosser, Washington, provided between 72 and 86 percent control of the root-knot nematode in that crop. In the same study, on-farm research in western Idaho showed that rapeseed green manures decreased soil populations of rootlesion nematodes to a greater extent than did sudangrass green manures. Fall sudangrass should be plowed down after it is

stressed. Winter rapeseed and canola should be incorporated in very early spring.

Rotation

The best rotation to control the Columbia root-knot nematode in potatoes involves planting a summer non-host crop, followed by a winter cover crop (rapeseed) incorporated as a green manure. Non-host crops include supersweet corn, pepper, lima bean, turnip, cowpea, muskmelon, watermelon, squash, rapeseed, canola, mustard, and sudangrass. For root lesion nematode control on potatoes, researchers found that forage pearl millet and marigold (Crakerjack) as rotation crops with potatoes resulted in fewer root lesion nematodes and increased potato yields than rotation with rye.

Marigold (*Tagetes species*) is one of the most highly studied crops for its ability to suppress nematodes with antagonistic phytochemical exudates, namely the polythienyls. Research also demonstrates that rhizobacteria living in association with marigold roots are suppressive to root lesion and other nematodes. These multiple effect nematode-control properties can benefit other crops when marigolds are grown in rotation. African marigold (*Tagetes erecta*) and French marigold (*Tagetes patula*) are popular ornamentals in the horticultural trade with several nematode-suppressive varieties each. Muster John Henry or little marigold (*Tagetes minuta*) is sold as the "Nematicidal" marigold, but it controls a relatively limited range of nematode species and readers should note that it is classified as a noxious weed in California. Tomatoes planted two weeks after African marigolds (*Tagetes erecta*) were disked into the soil showed a 99 percent reduction in root-lesion nematode damage compared to a tomato-tomato or fallow-tomato rotation. The French marigold cultivar "Single Gold" provided 99 percent control of nematodes in Dutch tests. Burpee Seed Co. has carried a French marigold variety known as "Nema-gone." The most effective marigold cultivars are those that germinate quickly, grow vigorously, and have deep root penetration. Cover crops exhibit tremendous variability in their susceptibility to or suppression of the four major types of plant-parasitic nematodes. For example, cover crops that suppress root-knot nematodes may be susceptible to sting nematodes. It is important to identify the nematode species in the field—and know what their plant hosts and antagonists are—before planning a cover-cropping strategy.

Allies from the Prairie

In Ontario, certain prairie species have been found to provide excellent nematode control when used as a cover crop, including blackeyed susan, gaillardia, and switchgrass, according to Marvin Pritts, PhD, of Cornell University. Another North American native known as "Indian Blanket" or "Blanket Flower" (*Gaillardia pulchella*) was effective in controlling southern root knot nematode (*Meloidogyne incognita*) on sweet potato. Tissue extracts of Indian Blanket were lethal to various plant-parasitic nematodes but were innocuous to free-living nematodes. Root exudates of Indian Blanket were lethal to mobile juvenals of *M. incognita* and were inhibitory to the hatch of eggs at concentrations of 250 parts per million or higher. Indian Blanket could be used to manage southern root knot nematode as a rotation crop, a co-planted crop, or a soil amendment to control root-knot nematode.

Fields left fallow but kept weed-free for one to two years usually have an 80 to 90 percent per-year reduction in root-knot populations. This host-free period can be achieved in one season, rather than two years, by disking every ten days all summer. While such disking offers the added advantage to reduce perennial weeds, it is expensive in terms of fuel costs, possible erosion, and loss of organic matter through oxidation.

Botanical Nematicides

Certain plants are able to kill or repel pests, disrupt their lifecycle, or discourage them from feeding. Some of these—marigolds, sesame, castorbean, and various brassicas—have been discussed previously as nematode-suppressive cover crops. In this section we will look at plants whose extracts or essential oils can be applied as nematicides.

For hundreds of years, Indian farmers have used the neem tree (*Azadirachta indica*) for its pesticidal, antifungal, and antifeedant properties. In research trials, potting soil amended with plant parts from the neem tree and Chinaberry tree (*Melia azadirach*) inhibited root-knot nematode development on tomatoes. However, no neem products are currently registered in the U.S. for use against nematodes. Margosan-O™, Azatin™, Superneem 4.5™, Neemix™, and Triact™ are neem products registered as insecticides, fungicides, and miticides. Neem cake, made from crushed neem seeds, provides nitrogen in a slow-release form in addition to protecting plants against parasitic nematodes. It is sold as a fertilizer in the U.S. through many farm and garden supply stores. Neem cake can be mixed with fertilizers such as composted manures, seaweed, and kelp. Recommended rates are 180 to 360 lbs. per acre or 2 lbs. per 100 to 160 sq. ft. Neem cake is toxic to plant-parasitic nematodes and is not as detrimental to beneficial free-living soil organisms. In greenhouse trials, 1 percent neem cake (mass/mass soil) caused a 67 to 90 percent reduction in the number of lesion (*Pratylenchus penetrans*) and root-knot (*Meloidogyne hapla*) nematodes in tomato roots grown in three different soils. In the field, 1 percent neem cake (mass/mass soil) reduced the number of lesion nematodes by 23 percent in corn roots and 70 percent in soil around roots.

Essential oils from various plants have shown promise as potential sources for new nematicides. Most of these plants are aromatic and culinary herbs that contain the nematicidal compounds carvacrol and thymol. At very low concentrations (1000 micrograms per litter, or .001 gm per litter, or .0038 gm per gal, or 0.38 gm per 100 gal) several oils immobilized juvenile root-knot nematodes and some also reduced hatching of eggs. The essential oils from the following plants ranked the highest for nematicidal activity: caraway, fennel, applemint, spearmint, Syrian oregano, and oregano. The toxicity of the essential oil from wormwood or Sweet Annie (*Artemisia annua*) leaves was evaluated in vitro against second-stage juveniles (J2) of the root knot nematode (*Meloidogyne incognita*) and pre-adults of the reniform nematode (*Rotylenchulus reniformis*). Complete mortality (100 percent) of both nematodes was found in 500 and 250 parts per million concentrations of the essential oil and gradually decreased with lower concentrations.

Biocontrols

Several microbial pathogens have been developed into commercial formulations against nematodes. These include the bacteria *Pasteuria penetrans* (formerly known as *Bacillus penetrans*), *Bacillus thuringiensis* (available in insecticidal formulations) and *Burkholderia cepacia*. Nematicidal fungi include *Trichoderma harzianum*, *Hirsutella rhossiliensis*, *Hirsutella minnesotensis*, *Verticillium chlamydosporum*, *Arthrobotrys dactyloides*, and *Paceilomyces lilacinus*. Another fungus, *Myrothecium verrucaria*, found to be highly effective in the control of nematodes, is available in a commercial formulation, DiTera™, from Abbott Laboratories. Circle One, Inc. offers a combination of several mycorrhizal fungal spores in a nematode-control product called Prosper-Nema™. Stein Microbial products offers the bacterium *Burkholderia cepacia* in a product called Deny™ and Blue Circle™. Rincon-Vitova offers a product called Activate™ whose active ingredient is the bacterium *Bacillus chitinosporus*.

The insect-attacking nematode *Steinernema riobravis* can provide root-knot nematode control comparable to that achieved with chemical nematicides. Although the exact mechanisms of control are not known, researchers hypothesize that an allelochemical is involved (perhaps manufactured by symbiotic bacteria that live within *S. riobravis*) that repels plantparasitic nematodes. Recent research measured the effect of beneficial nematodes on root-knot nematodes (*Meloidogyne species*) infecting tomatoes and peanuts. In the laboratory, peanut seedlings treated with the beneficial nematodes *Steinernema feltiae* and *Steinernema riobrave* showed resistance to pest nematodes. In the greenhouse, scientists tested application levels and timing on peanut and tomato plants. On peanuts, pre-and post-infestation applications of *S. feltiae* suppressed *M. hapla* penetration but not egg production. Only pre-infestation applications of *S. riobrave* suppressed *M. hapla*. The tomatoes were infested with *Meloidogyne incognita* eggs and treated with *Steinernema glaseri* or *Heterorhabditis megidis* applied at the same times as the tomato treatments. The low rate of *S. glaseri* suppressed *M. incognita* penetration into tomato roots and the high rate of *S. glaseri* reduced egg production. Those interested in using this biocontrol will need to experiment with application rates and techniques to develop methods best suited to their operations. Additional information on insect parasitic nematodes can be found on an Ohio State University Web site.

A soil-dwelling predatory mite, *Hypoaspis miles*, preys primarily on fungus-gnat larvae but will also attack spring tails, thrips, and nematodes. (Anon, No date) These mites are available commercially for the control of fungus gnats in greenhouse production of tomatoes, peppers, cucumbers, flowers, and foliage plants. The mites are applied to the planting media.

It is clear that a wide range of organisms feed on, kill, or repel nematodes. These organisms are most effective, and are found most commonly, in healthy, well-managed soils.

Plant Resistance

Generally speaking, a resistant cultivar is more effective against sedentary endoparasitic species such as root-knot and cyst nematodes than against "grazing" ectoparasitic species. Root-knot and cyst nematodes spend most of their lifecycle within the root, relying on specialized cells for feeding. Upon entering the roots of resistant cultivars, these nematodes become trapped as the feeding cells necessary for their survival fail to develop.

Many crop cultivars—tomatoes and soybeans in particular—have been specifically bred for nematode resistance. The "N" designation on tomato seed packages (usually as part of "VFN") refers to nematode resistance. A few cultivars of potatoes are resistant to the golden nematode, which is a pest only in a small area of the northeastern U.S. Although most cultivars of potatoes are susceptible to infection by nematodes, some varieties tolerate infection better than others. For example, population densities of root-lesion nematodes (*Pratylenchus penetrans*) that would affect yield in "Superior" are tolerated with little effect by "Russet Burbank."

Richard L. Fery, PhD, a geneticist at the USDA's Agricultural Re search Service in Charleston, South Carolina, developed two nematode-resistant varieties of bell pepper, "Charleston Belle" and "Carolina Wonder," available from commercial seed companies. Charleston Belle and its susceptible parent, "Keystone Resistant Giant," were compared as spring crops to manage the southern root-knot nematode (*Meloidogyne incognita*) in autumn-cropped cucumber and squash. Cucumber grown in plots following Charleston Belle had lower root gall severity indices than in crops following Keystone Resistant Giant. Cucumber yields were 87 percent heavier and numbers of fruit 85 percent higher in plots previously planted to Charleston Belle than to Keystone Resistant Giant. Squash grown in plots following Charleston Belle had lower root gall severity indices than those following Keystone Resistant Giant. Squash yields were 55 percent heavier and numbers of fruit 50 percent higher in plots previously planted to Charleston Belle than to Keystone Resistant Giant.

These results demonstrate that root-knot nematode-resistant bell pepper cultivars such as Charleston Belle are useful tools to manage *M. incognita* in double-cropping systems with cucurbit crops. Nematode-tolerant or resistant cultivars of snap beans, lima beans ("Nemagreen"), and sweet potatoes ("Carolina Bunch," "Excel," "Jewel," "Regal," "Nugget," and "Carver") also exist and may be used in a similar strategy to reduce nematode levels for crops that follow.

The choice of nematode-resistant rootstock for perennial fruit production is important to ensure protection of trees and vines against these unseen pests. Consult with a local farm advisor to confirm that the rootstock you choose is appropriate for the area.

Table: *Nematode-resistant rootstock for perennial fruit.*

Fruit	***Rootstock***
Apple	No commonly used rootstock is completely resistant
Pears	Bartlett, Quince (slight resistance)
Asian Pear	Calleryana
Citrus	Poncirus trifoliate, lime, rough lemon, sour orange Forner-Alcaide 5
Grapes	Freedom, Harmony, Dog Ridge, Ramsey VR039-16
Peach & Nectarines	Nemaguard, Nemared, Citation, Hansen 536
Plums	Myrobalan 29-C, Marianna 2624
Apricots & Almonds	Nemaguard, Nemared, Myrobalan, Marianna 2624
Cherries	Mazzard, Mahaleb

Breeding for nematode resistance in most crops is complicated by the ability of the nematode species (primarily cyst nematodes and root-knot nematodes) to develop races or biotypes that overcome the genetic resistance factors in the crop. In order to maintain resistant crop cultivars on farms, researchers suggest that susceptible and resistant cultivars be planted in rotation. When a nematode-resistant cultivar is planted, nematode populations generally decrease, but over the course of the growing season the few nematodes in a particular population capable of overcoming this resistance begin to increase. If in the following season the farmer plants a susceptible cultivar, the overall nematode numbers will still be low enough to avoid significant yield reduction, but more importantly, the selective pressure favouring the increase of the "counter-resistant" bio types is removed. As long as the farmer continues to alternate susceptible and resistant cultivars (and, better yet, incorporate non-host crops into the rotation), the nematodes can be kept at non-damaging levels.

Red Plastic Mulch

Springtime field tests at the Agricultural Research Service in Florence, South Carolina, indicate that red plastic mulch suppresses root-knot nematode damage in tomatoes. According to Michael Kasperbauer, one of the researchers, "The red mulch reflects wavelengths of light that cause the plant to keep more growth above ground, which results in greater yield. Meanwhile, the plant is putting less energy into its root system—the very food the nematodes feed on. So reflection from the red mulch, in effect, tugs food away from the nematodes that are trying to draw nutrients from the roots." The research team planted tomatoes in sterilized soil, mulched them with red or black plastic, and inoculated the roots with nematodes. Plants inoculated with 200,000 nematode eggs and mulched with black plastic produced 8 pounds of tomatoes, while those mulched with red plastic produced 17 pounds. The red mulch is available commercially from Ken-Bar, Inc., of Reading, Massachusetts.

Solarization

Soil solarization, a method of pasteurization, can effectively suppress most species of nematode. However, it is consistently effective only where summers are predictably sunny and warm. The basic technique entails laying clear plastic over tilled, moistened soil for approximately six to eight weeks. Solar heat is trapped by the plastic, raising the soil temperature. The incorporation of poultry litter prior to solarization, or use of a second layer of clear plastic, can reduce effective solarization time to 30 days. Brassica residues are also known to increase the solarization effect, in a process known as biofumigation. The plastic holds in the gaseous breakdown products of the brassica crop (or food processing wastes), thereby increasing the fumigation-like effect. Large-scale field experiments using cabbage residues with solarization obtained results comparable to solarization combined with methyl bromide. Solarization is well documented as an appropriate technology for control of soilborne pathogens and nematodes, but the economics of purchasing and applying plastic restrict its use to high-value crops.

Soil Steaming

Steaming the soil suppresses nematodes in a manner similar to solarization. There are prototype steam machines capable of performing field applications, but steaming is probably economical only for greenhouse operations or small plantings of high-value crops.

Flooding

In certain parts of the country where water is usually available and water pumping equipment and dikes already exist, and for certain large-scale monocultures (e.g., potatoes), flooding is sometimes used as a management tool to control nematodes. But for most farms, it is probably not an option. Flooding the soil for seven to nine months kills nematodes by reducing the amount of oxygen available for respiration and increasing concentrations of naturally occurring substances—such as organic acids, methane, and hydrogen sulfide—that are toxic to nematodes. However, it may take two years to kill all the nematode egg masses. Flooding works best if both soil and air temperatures remain warm. An alternative to continuous flooding is several cycles of flooding (minimum two weeks) alternating with drying and disking. But note that insufficient or poorly managed flooding can make matters worse, as water is also an excellent means of nematode dispersal.

Summary

Each combination of nematode and host is different. As the nematode population density reaches a certain level, the host crop yield suffers. Some hosts support faster population increases than others. Environmental conditions can also affect the relative dangers posed by nematode populations. As we begin to develop a better understanding of the complex ecologies of soils and agricultural ecosystems, more strategies for cultural and biological control of nematodes will be developed. The trick will be fine-tuning these general strategies to the unique ecology, equipment, and financial situation of each farm.

Opportunities for Integrated Management of Plant-parasitic Nematodes in the Near East

Plant-parasitic nematodes must be addressed in crop production and integrated pest management (IPM) systems if agriculture is to meet the world demands for increasing food and fibre production. On a worldwide basis, annual crop losses due to nematode damage have been estimated to average 12.3 percent, amounting to some US$77 million annually. The data available indicate that similar losses occur in the Near East. For example, estimated losses for vegetable crops due to nematode-related disease complexes in Egypt amounted to some 15 percent in 1986, with losses for field crops ranging from 5 to 20 percent. Nevertheless, the subtle nematode symptoms and signs are often confused with nutrient deficiencies and other maladies, resulting in nematodes being overlooked by agricultural scientists as well as growers.

The concept of combining compatible tactics for controlling nematodes predates that of IPM. In 1889, Atkinson discussed a range of nematode-management tactics that is surprisingly similar to those available today. Tactics discussed for controlling root-knot nematodes include: sterilization of the soil by starvation (including the use of non-host plants); the potential of trap crops; composts; nematicides; and soil amendments such as hardwood ash and potash. The early work of Cobb (1918) on sampling nematode communities provided a basis for the development of improved tactics and strategies essential for integrated pest management. The following statement was offered by Tyler in 1933:

"A well-planned combination of practices will go much further for control of nematodes than any of the recommended treatments alone". Tyler also stated that the presence of nematodes (root-knot) in any soil can best be determined by examining the roots of susceptible plants which have been growing for at least three weeks in warm, moist conditions. For attempted eradication of root-knot nematodes, the author suggested the following:

1. Burning of residues two or three times if possible, each preceded by digging or ploughing.
2. Dry fallow with frequent ploughing.
3. One or two well-irrigated trap crops, completely destroyed two or three weeks after sprouting.
4. Moist fallow during warm weather, without weeds.
5. Resistant crops in rotation, kept free of weeds.
6. Repetition of steps 4 and 5.

Tyler also emphasized that root-knot nematodes enhance the susceptibility of crops to other diseases such as cotton wilt, black shank of tobacco and *Rhizoctonia* disease of groundnuts. Thus, early nematologists provided an exceptionally strong conceptual framework for the development of advanced integrated management systems.

The specific concept of IPM is a relatively recent development in pest control. The 1962 publication of *Silent Spring* by Rachel Carlson and the subsequent banning of DDT awakened United States agriculturists, especially entomologists, to the risks of pesticides and provided the impetus for the development of IPM. IPM consists of the development, use and deployment of pest-control strategies and tactics that result in favourable socioeconomic and environmental consequences. Theoretically, IPM is a holistic systems approach to limit pest damage to tolerable levels through a combination of tactics and techniques, including parasites and predators, host resistance, cultural practices, environmental modification, and pesticides where appropriate. Unfortunately, the interest and progress of IPM programmes has not been balanced among the pest disciplines with the major emphasis often being on insects. Still, a number of national and international IPM programmes have resulted in many research and extension programmes becoming more interdisciplinary.

The relative youth of the science of nematology and the availability of effective, relatively inexpensive nematicides hampered the development of integrated management systems in this discipline. During the last two decades, however, much progress has been made in characterizing damage functions and thresholds for major nematode-crop combinations, the study of general population dynamics and simulation models for key nematodes, and the development of improved management tactics. In addition, considerable progress has been made in elucidating the roles of nematodes in disease/pest complexes, and their interactions with symbionts.

Components of integrated pest management

IPM has been divided into a number of components in an operational framework including:

- biological monitoring;
- environmental monitoring;
- the decision-maker;
- the decision-support system;
- the decision;
- procedure implementation;
- the system.

Biological Monitoring

Dispersal patterns of plant-parasitic nematodes over space and time must be considered in characterizing populations for IPM purposes. Contagious or spotty horizontal dispersal patterns in the field pose difficulties in sampling, experimentation and management. Vertical dispersal patterns also vary greatly, but these obligate parasites are generally closely associated with host roots. Major factors that influence the seasonal fluctuations of nematode populations include: their biology (especially reproductive and survival characteristics); environmental parameters; and, especially, management practices. The temporal population dynamics of *Meloidogyne incognita* are typical of many nematodes. This pathogen increases to very high population densities during the growing season and then declines very abruptly after harvest.

This decline, however, may be greatly affected by management practices. Note, in figure below, the striking decline of *M. incognita* numbers in plots in which residual roots were destroyed by repeated discing, in contrast to the relatively high survival in the absence of ploughing. Frequently, these declines are so severe that the population levels may become non-detectable by routine sampling.

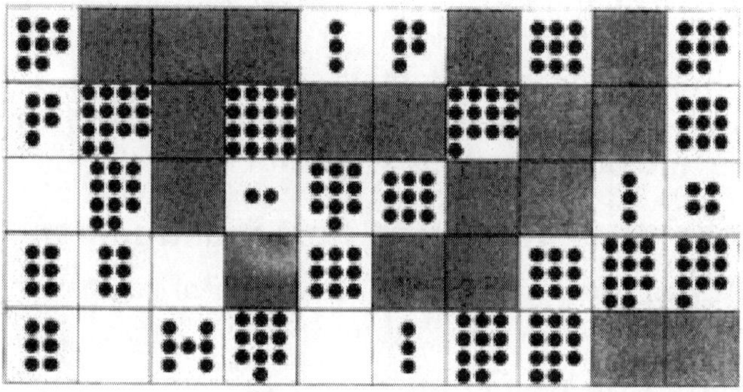

Figure: *Contagious dispersal pattern of juveniles of* Meloidogyne incognita *(spring samples). Each dot in the 5 × 5 m quadrants represents 25 juveniles per 500 cm³ of soil; dark quadrants had numbers of 400 or greater for same soil volume. Computer programs are now available for this type of mapping*

Advice for nematode IPM programmes may be based on qualitative and/or quantitative information. Today, many management decisions are based on qualitative or, at best, semi-quantitative population thresholds. This problem is related to the low level of precision in sampling schemes, limited information on damage functions, and the effects of crop cultivar and environmental parameters on nematode/host interactions. Despite these limitations, nematode monitoring can offer much to IPM programmes for growers.

For example, the detection of small numbers of a highly damaging species, such as *Meloidogyne arenaria* on groundnuts, may require a shift to another crop or a change in management tactics. Large numbers of a cyst nematode such as *Heterodera zeae* may require long-term rotations or a combination of management tactics.

As suggested by Tyler in 1933, bioassays utilizing suitable host plants have much to offer in managing certain nematodes. This approach has been used successfully for *Heterodera avenae* on wheat in Australia. Critically timed gall ratings of crop roots, and/or North Carolina differential host plants, are very useful for identifying species and host races of *Meloidogyne*. This approach to monitoring nematode populations can be adapted readily for grower use. Carefully and properly timed compilations of root-gall and root-necrosis indices also are useful as a basis for obtaining yield-loss estimates from these pathogens.

Much progress is being made in the development of new technologies for reliable nematode diagnostics and identification which are critical to IPM. Biochemical techniques offer promise to nematode identification and possibly population assessment.

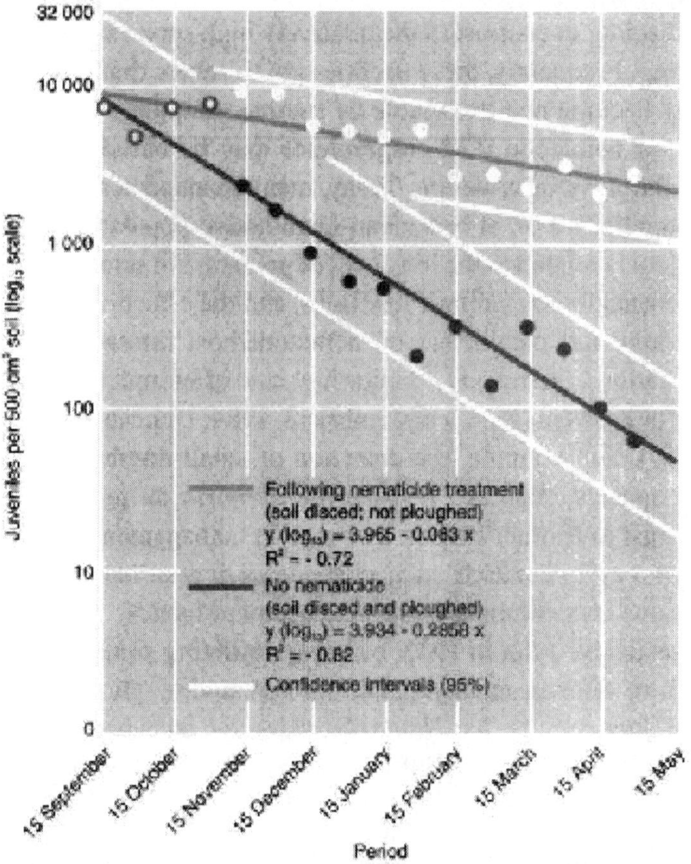

Figure: Effects of management practices on survival of Meloidogyne incognita after a tobacco crop (nematodes extracted by centrifugal flotation)

Environmental Monitoring

Nematode population dynamics and associated crop responses are often related to environmental parameters. Thus, such information is critical to effective IPM nematode programmes. In addition to enhancing the predictability of crop responses to given nematode species, environmental data are also helpful in achieving optimum efficacy of a number of management tactics. For example, data on expected soil temperatures should prove useful for maximizing the efficiency of solarization, bare fallow, or combinations with tillage and/ or residual root destruction.

Decision-making

The focal point of IPM programmes should be on the decision-maker. Thus, the decision-maker should have an understanding of nematode biology, the system to be managed and possible effects of IPM practices on the environment. Although only recently introduced to plant pathology and nematology, expert systems have much potential in this regard. For an IPM system to be successful, it must have a decision-support system. The minimum support should include innovative research programmes and effective information delivery/transfer. Support systems must address related socioeconomic issues as well as nematode damage functions/ thresholds, sampling, management options, education and grower inputs. Computer communications and simulation models are becoming increasingly important in IPM and nematology.

Management decisions are based on assessment of the problem, related economics, and availability of a range of management strategies and tactics. (Factors affecting the efficacy of given tactics will be discussed with the respective options.) In management decisions, it is important that compatible tactics be selected and that they be utilized in a proper, effective application. The key topic of nematode-damage functions and thresholds is discussed by Dr Trudgill on page 45. Where damage thresholds and related information are unavailable, nematode-hazard indices (relative probability of moderate to severe damage) might be considered.

Reassessment of the system through biological and environmental monitoring is the final step in IPM. For nematodes, this should include follow-up population assessments to determine the effectiveness of the management tactics deployed. Long-term monitoring and assessments of given cropping systems should facilitate the detection of shifts in nematode communities, including new species, races, or resistance-attacking biotypes.

Overall IPM Constraints

The general constraints to developing effective IPM programmes vary with each country or region. Major overall constraints common to most countries concern technology development and transfer, institutional resources and related training, environmental factors, and key economic and socio-cultural components. For example, growers are often excluded in the early phases of IPM development, yet they may have much to offer and their support is essential for a successful programme. This list of constraints implies that IPM requires interdisciplinary research and extension efforts. This truly integrated approach to IPM has been achieved in too few instances to date. Nevertheless this issue is central to effective pest management because of the many pest interactions and pesticide interactions.

Nematode Management, Strategies and Tactics

The terms strategy and tactics have been used interchangeably in some instances in nematology. In this discussion, Vanderplank's (1963) usage of these concepts will be followed, that is: strategy refers to the overall plan for pathogen (or nematode) management, and tactics denote the specific tools or weapons utilized to carry out the plans. Thus, the general strategies for managing nematodes include:

1. Exclusion (quarantine).
2. Reduction of initial inoculum density.
3. Suppression of nematode reproduction.
4. Restriction of damage to the current crop.

Some researchers combine strategies 2 and 3 under population density management (Bird, 1987). The specific tactics and relative efficacy of the respective strategies are outlined in table.

Currently, most nematode management programmes focus on tactics involving the reduction of the initial nematode population density and/or the suppression of their reproduction during the season. Most nematode management tactics also have the inherent feature of limiting damage associated with the target crop. In contrast, tolerant crop cultivars may be damaged little economically while supporting relatively high nematode reproduction. With the diminishing availability of a single strategy or tactic for nematode management, a combination of two or more compatible tactics in an integrated system is becoming more critical than in recent years when nematicide usage was the primary measure for nematode control.

Exclusion or Avoidance

Fortunately nematodes have a number of characteristics that limit their long-distance dispersal, including their restricted active movement, obligate parasitism, narrow host ranges of many species, sharp population declines in the absence of hosts, and survival depending on environmental conditions and crop management practices. Thus, long-distance dispersal of these parasites is largely passive and by chance. Key means of dissemination include movement of soil on equipment and plant parts, crop transplants, water, animals and contaminated containers such as burlap bags. Measures for avoiding dissemination and establishment of new nematode problems should be a component of national or regional nematode IPM programmes.

Table: *Relative efficacy of nematode management strategies and tactics*[1]

Strategy[2]	*Tactic Relative efficacy*[3]	
1. Exclusion/avoidance	Quarantine	M,H
2. Reduce initial population density	a. Eliminate established foci	L,M
	b. Cultural	
	• use of clean planting stock	H
	• crop rotation	L-H
	• inter-and intracropping	L,M
	• cover/trap crops	L,M
	• soil amendments	L,M
	• fallow	L-H
	• time of planting/harvesting	L,M
	• farm hygiene and general culture	M,H
	c. Vertical resistance	H
	d. Chemical nematicides[4]	
	• fumigants	M,H
	• non-fumigants	L-H
	e. Biological	
	• natural	L-H

Contd...

Strategy[2]	***Tactic Relative efficacy***[3]	
	• introduced, including enineered	L-H
	f. Physical	
	• tillage, including residual root destruction	L-H
	• heat solarization	L,M
	• flooding	L,M
3. Suppress nematode reproduction	a. Resistance horizontal/quantitativeb. Protection with supplemental chemical nematicides	L,M L,M
4. Restrict current crop damage	a. Tolerant cultivarsb. One or more of tactics under 1 and 2	L-H

[1] In part after Oostenbrink, 1972 and Roberts, 1993.

[2] Strategies No. 2 and 3 after Vanderplank, 1963.

[3] Low (L); moderate (M); high (H).

[4] Chemical nematicides often result in high carryover population densities.

Government-regulated quarantines have had varying levels of success in different countries. For example, the quarantines for *Globodera rostochiensis* and *Heterodera zeae* have been success stories to date in the United States. In contrast, the quarantine for the soybean-cyst nematode, *Heterodera glycines,* failed in that country. This failure may have been a result of the nematodes being more widely disseminated than initially realized as well as having several effective means of dispersal. Internal quarantines developed by the European and Mediterranean Plant Protection Organization have focused on *Aphelenchoides besseyi* for rice and strawberry, *Ditylenchus destructor, G. rostochiensis* and *G. pallida* on potato, *Radopholus citrophilus* on citrus, and *Xiphinema americanum* on a range of crops. More recently, *Bursaphelenchus xylophilis* has been a focal point concerning the importation of wood products in a number of countries. Other nematodes that have limited distribution and are threats to certain crops include *Meloidogyne chitwoodi* and *M. nataliei*. Further spread of these and other highly aggressive nematodes is undoubtedly a concern in most Near East countries.

At a more practical level, certified plant material, nematode-free planting stock and clean farm equipment should be standard

IPM practices. This issue is particularly important for vegetatively propagated crops such as banana, potato, and crops for which seedlings are transplanted. *In vitro* propagation of banana has proved to be an effective means of providing clean planting stock and should reduce the need for nematicides.

In addition, a number of nematodes are disseminated with crop seeds, including *Anguina tritici* on wheat *and Ditylenchus dipsaci* on alfalfa and a number of other crops.

Reduction of Initial Nematode Population Densities

Six general tactic categories may be placed under this strategy: crop eradication; rotation; vertical resistance; chemical nematicides; biological control; and physical treatments to reduce nematode population densities. This strategy includes some of the oldest and most recently developed tactics that have been used in carefully developed IPM systems.

Eradication or containment. Although eradication is a longstanding tactic for nematode management, it is very difficult to achieve success for field infestations. Fumigation or heat treatments, for example, may eliminate all nematodes in upper soil layers, but those occurring at depths of 0.3 to 1.0 m below the soil surface will generally escape and eventually move back into the plough layer. In contrast, nematode infestations in confined environments, including greenhouses and container-propagated plants, can be eliminated through standard sanitation procedures.

Cultural practices. A wide range of cultural practices have varying levels of efficacy in nematode management. These include clean planting stock, crop rotation, inter-and intracropping, cover/trap crops, soil amendments, fallow, time of planting/harvesting, and general farm hygiene and culture. Such non-chemical tactics are especially important for low-value crops.

Crop rotation. This is one of the oldest and most effective tactics for managing plant-parasitic nematodes. The goal of rotation is to bring about a striking population decline of the target nematodes that will facilitate the subsequent crop to grow and produce an acceptable yield. Crop rotation and cropping systems are similar concepts which are sometimes confused. Crop rotation is the fixed yearly sequence and spatial arrangement of crops, or the alternation of crops and fallow, on a given land area. Alternate crops in rotation may be planted or natural. In contrast, a cropping system is the sequence of growing crops and the required technologies for their production. Thus, cropping systems cover all kinds of crop sequences, including monoculture, whereas crop rotation indicates an inflexible cycle or a fixed sequence of crops.

The study of cropping systems should include quantitative analysis of the relationships among crops and pests, and general management tactics which may be deployed in that system. The crop system may be spatial or temporal. Thus nematode populations may respond to individual crops as well as the arrangement of the crops over space and time. To maximize the efficacy of rotation and cropping systems in nematode management, the short-and long-term effects of cropping systems, spacing and associated interactions with biotic and abiotic environmental factors on crop yields must be better understood. Research addressing these issues should include potential effects of the general environment, associated weeds and other plant pests in cropping systems and development of appropriate crop rotations.

Nusbaum and Ferris (1973) gave two key principles for crop rotation: the reduction of initial nematode population levels necessary to permit the subsequent crop to complete its growth before severe attack; and the preservation of competitive, antagonistic and/or predacious nematodes and other organisms at population densities that allow effective buffering of the pathogenic species. Shifts in nematode populations and communities may be characterized numerically through the application of the concepts that define host status and by various population models. For example, good hosts have a high equilibrium density (E) and maximum rates of reproduction (a) (Seinhorst, 1970; Nusbaum and Ferris, 1973). In contrast, these parameters would be low for poor hosts or non-hosts. The relative host sensitivity is defined or indicated by the tolerance limit (T) (Seinhorst, 1970). The utilization of these quantitative concepts for target nematode species for a primary host and associated crops is necessary for the development of effective rotation systems.

Table: *Types of crops and their arrangement in time and space evaluated with reference to potential development of pest problems. Some effects are high in pest potential. some intermediate and some low*[1]

High pest potential	**Low pest potential**
General types of crops and associated pest problems	
Large pest complex ∅	Small pest complex
(crop not competitive with weeds)	(crop highly competitive with weeds)
Susceptible ∅ Tolerant cultivar ∅	Resistant pure line ∅ Resistant multigenic
Annual ∅	Perennial
Long-maturing∅	Short-maturing
Crop arrangement in time	
Monoculture∅	Crop species rotation
Continuous planting∅	Discontinuous planting
Asynchronous planting∅	Synchronous planting
Season favourable to pest∅	Season unfavourable to pest
Crop arrangement in space	
Sole cropping∅ Row or strip∅ intercropping	Mixed intercropping
High plant density∅	Low plant density
Large field∅	Small field
Large host crop area∅	Small host crop area
Host fields aggregated∅	Host fields scattered

[1] *Partially after Raymundo (1985) and Trivedi and Barker, 1986.*

Much knowledge is essential for quantifying nematode-host interactions and the development of specific recommendations for crop-rotation systems. Such data must include nematode species and race or biotype, their host ranges, host efficiency and susceptibility of crops to be included in the rotation, associated weeds, general population dynamics, and information on the relationships between population densities and crop loss. Other critical information includes the effects of abiotic environmental parameters on nematode population dynamics and associated disease complexes and possible pesticide interactions.

The level of success in nematode management through crop rotation may vary with the year, target nematodes and associated plant pathogens, location, weed hosts present, and the nature and length of the rotation. Publications that give host ranges, including weeds, for various nematodes are available. Details on the host ranges of given nematodes are beyond the scope of this discussion. Nevertheless, the non-hosts and poor hosts tabulated for given nematodes, may serve as a starting point for developing effective crop rotations for IPM purposes.

Table: *List of selected non-host plants useful in crop rotation*[1]

Nematode species	*Non-hosts or poor hosts*[2]
Belonolaimus longicaudatus	*Crotalaria* spp., *Crotalaria spectabilis* hairy indigo, marigold, tobacco
B. gracilis	*Crotalaria spp.*, tobacco, watermelon
Dolichodorus heterocephalus	Crotalaria spectabilis
Helicotylenchus dihystera	Alfalfa, maize, fescue[2]
Heterodera glycines	Maize, cotton, cowpea, potato, small grains, grains, tobacco, most vegetables
H. schachtii	*Alfalfa, bean, clover, maize, Hesperis matronalis,* onion
H. zeae[3]	Wide range of crops
Globodera rostochiensis	Maize, green beans, red clover
Hoplolaimus indicus	Cabbage, chili, eggplant
Meloidogyne javanica	*Andropogon, Crotalaria* spp., cotton, groudnut, sorghum, velvet bean
M. hapla	Maize, cotton, grasses, lettuce, onion, radish
M. incognita	Fescue, orchard grass
Meloidogyne spp.	*Crotalaria spectabilis, Indigolera hirsuta,* millet, oats, wheat[2]
Paratrichodorus minor	Maize,[2] Crotalaria spectabilis
Pratylenchus leiocephalus	Groundnut
P. penetrans	Alfalfa,[2] beet, fescue, marigold, oats, Sudan grass, rye
Pratylenchus spp.	Lettuce, onion, radish
Radopholus similis	Crotalaria spectabilis
Tylenchorhynchus mirzal	Wheat
T. brassicae	Potato, tomato
Xiphinema americanum	Alfalfa,[2] maize,[2] fescue,[2] tobacco

[1] *See Trivedi and Barker (19S6) for specific references per nematode/crop combination.*

[2] *Some populations of respective nematode species will reproduce rapidly on these crop plants.*

[3] *Host range needs further study.*

Non-host crops have been used for developing effective rotation programmes for nematodes such as *Meloidogyne* spp., *Heterodera* spp., *Ditylenchus* spp., *Belonolaimus* spp. and *Pratylenchus* spp.. Rotation systems that involve two to four years of non-host crops generally give excellent results managing these nematodes. For cropping systems in the Near East, this may involve several crop sequences of non-hosts between the production of a susceptible host during such a two-to four-year period. For example, groundnuts, cotton, potato and a range of vegetables or perennial crops, such as alfalfa, may be grown between crops of maize as a means of limiting damage to the latter by *H. zeae*. Effective crop rotations for *Meloidogyne* spp., with their broad host range, are more difficult to develop. However, a number of crops important in the Near East may be considered as a nucleus for rotations involving species of *Meloidogyne*. Cotton is a non-host for *M. arenaria* which attacks groundnut, and the latter is a non-host for *M. incognita* which attacks cotton. Some cultivars of wheat are also poor hosts for both of these nematodes. Thus, such limited information can be used as a starting-point for rotation systems for these important crops where sufficient land is available. Much information on the distribution of major nematodes, their host ranges and potential management options, including rotation, is becoming available for Near East countries. Also, more comprehensive data on potential weed hosts are becoming available.

Inter-and intracropping is a widely used practice in several countries. Various types of mixed cropping and asynchronous temporal planting may limit nematode damage as outlined by Raymundo (1985). Definitive data on these practices are rather limited.

Cover/trap crops and antagonistic plants are useful for reducing nematode populations as well as conserving soil and often improving soil texture. In localities where land availability permits, the use of cover crops, especially plants that serve as trap crops or offer other suppressive effects on nematode populations, should be considered. Carefully selected cover crops may serve as living mulches and provide multiple pest control.

Table: *Trap crops for plant-parasitic nematodes*[1]

Nematode species	***Trap crop***
Meloidogyne spp.	*Crotalaria spectabilis,* cowpea, English pea, periwinkle
Heterodera avenae	Oat
H. schachtii	Hesperis matronalis
Globodera spp.	Potato

[1] *In pan, after Trivedi and Barker (1986).*

Organic amendments offer considerable potential for restricting populations of plant-parasitic nematodes. Soil amendments with oilcakes, green manure, mature crop residues, chitin-containing materials, hemicellulose and farm manures may provide varying levels of nematode suppression. Leaves of neem (*Azadirachta* spp.) and products thereof provide nematicidal and other pesticidal activity. A chitin-based preparation was recently released as a nematicide in the United States. Unfortunately, this product failed to provide acceptable levels of nematode suppression. A combination of green manure crops or other organic amendments with promising biological control agents may have potential for nematode management on a large-field basis.

Fallow is a simple tactic for reducing nematode populations through starvation, as suggested by Tyler in 1933. The goal with this practice is to free the soil of all vegetation for varying periods of time, often by frequent tillage. Theoretically this tactic should be more effective in hot, arid climates then in cooler regions that have high rainfall. However, a number of nematode species inhabiting dry habitats may survive for extended periods through anhydrobiosis. Also, cysts of *Heterodera* or *Globodera* spp. may be tolerant to desiccation and/or high temperatures and survive for prolonged periods in the absence of host plants. Additional disadvantages of this tactic include the problems of increased soil erosion, adverse effects on soil structure, and the fact that this practice fails to contribute directly to farm income.

The time of planting/harvesting may be utilized to exploit differential environmental effects on nematode populations versus crop growth and maturity. For example, early planting of crops such as wheat, barley, rye, chickpea and potato has restricted associated nematode damage in some instances. Because of the prevailing temperatures and the conditions required for optimum growth of most crops, this approach often is impractical.

Overall farm hygiene and culture may also play a key role in limiting crop damage caused by plant-parasitic nematodes. The typical nutrient deficiencies resulting from root damage and impaired uptake effected by nematode parasitism may be offset, in part, by proper nutrition, moisture and protection from adverse conditions. Adequate potassium levels may enable a number of plants to tolerate large populations of nematodes. Ammonium sources of nitrogen have been shown to suppress a number of nematodes, including H. *glycines*. Based on research with D. *dipsaci* on alfalfa, high calcium levels may increase host resistance to nematodes. The integration of crop management and compatible cultural tactics might be viewed as the basic foundation for successful nematode control and crop production.

Vertical resistance. This type of resistance in plants to pathogens, including nematodes, theoretically falls under the strategy of reducing initial inoculum level. This situation clearly is true for resistant cultivars that function in large part as trap crops.

Chemical nematicides. Nematicides available on the world markets may be classified as fumigants and non-fumigants, and considerable research is focusing on the development of novel compounds. The properties that are responsible for soil fumigants being excellent nematicides may result in their being placed under regulatory review. To date, DBCP, EDP and DD have been withdrawn from United States markets through the action of regulatory agencies. In addition, 1,3-D has been removed from current use in California, and there is a risk that methyl bromide will be removed from the United States markets. The latter compound has been implicated as an agent that damages the ozone layer. Although a range of non-fumigant nematicides have acceptable levels of efficacy, most of these compounds have been detected in groundwater. In addition, some reports have focused on the microbial breakdown of some of these materials in soil, especially under continuous usage. In spite of the problems associated with nematicides, they comprise an important tactic for many nematode IPM systems, especially for high-value crops. As indicated by Sasser (1989), the procedure for selecting a nematicide should include procurement of information on efficacy and capacity to enhance yields of the primary crop or crops in a specific region. Sasser (1989) and Thomason (1987) provide details on nematicide safety and environmental issues, related economics and additional problems concerning factors in selecting and using these materials, their methods of application and dosage and use on annual or perennial crops.

Biological control agents. Increasing interest in this important nematode management tactic is reflected in two recent books. As we gain greater understanding of the ecology and biology of nematode antagonists, parasites and predators, they undoubtedly will become important in an integrated approach to managing plant-parasitic nematodes as suggested by Stirling (1991).

Physical. The concept of physical means of reducing nematode populations was considered by Atkinson (1889), Bessey (1911) and Tyler (1933). Postharvest ploughing or discing can result in a 90 percent or greater reduction of nematode populations. The prevailing high temperatures in much of the Near East may add to the effectiveness of this tactic during much of the year.

Heat solarization may be an effective nematode management tactic for small land plots in some regions of the Near East. For example, Saleh, Abu-Gharbieh and Al-Banna (1989) showed that a combination of the application of solar-heated water and solarization (with a plastic cover) resulted in a decline of 96 percent in populations of *Meloidogyne javanica* in Jordan. With this treatment, soil temperatures at 10 to 20 cm depths reached 60° and 56°C, respectively, whereas a maximum temperature of 46°C was achieved with solarization alone. Still, the addition of a nematicide or another tactic may be needed to achieve this level of control in some regions. The likely differential sensitivity of different nematode species to high temperatures should be considered in using this tactic. For example, *M. javanica* can tolerate temperatures of 50°C up to three hours or more, whereas *M. hapla* and *M. incognita* are killed in much shorter times at this temperature. Solarization also provides some multiple pest control.

Flooding has been evaluated periodically for decades as a means of reducing nematode populations. At high temperatures, and especially with organic amendments, this treatment can be effective in reducing nematode levels. In contrast, flooding plants such as cranberries under extremely cool conditions may have little impact on parasitic nematodes. Because of the extensive use of irrigation in much of the Near East, the potential of delivery of nematode antagonists or parasites with the water warrants careful study.

Suppression of Nematode Reproduction

Host-plant resistance will provide an increasingly important contribution to the solution of many nematode problems. With the enhanced availability of germplasm that has nematode resistance genes and rapidly advancing molecular-transfer techniques, resistant cultivars should become a primary management tactic in nematode IPM. This approach will become more critical as chemical nematicides become more restricted. The striking returns from deployment of some nematode-resistant cultivars reflect the great potential of this management tactic. For example, Bradley and Duffey (1982) in an economic analysis concluded that the soybean cultivar Forrest, resistant to *M. incognita* and *H. glycines,* gave a US$400 million return over a six-year period (the cost of development of this cultivar was US$1 million).

A number of reviews on the development and deployment of host-plant resistance are available. Except for the stem nematode *D. dipsaci,* most of the available host resistance is limited to species that have sedentary feeding habits. This situation probably results from three factors: the reproductive activity of these sedentary nematodes can be assessed more readily than migratory endoparasites or ectoparasites; the intimate host-parasite relationship of sedentary nematodes is more likely to be controlled genetically than those of migratory species; and the real or perceived importance of the maladies induced by sedentary nematodes results in much greater research on host resistance than for other feeding groups.

The list of highly effective host-plant resistances to given nematodes is rather limited. This listing is restricted to only seven nematode genera on a relatively narrow range of crops. Still, some resistance has been developed for other nematodes, including *Pratylenchus* spp., *Naccobus aberrans, Rotylenchulus reniformis, Radopholus citrophilus, Xiphinema americanum* and *Criconemella xenoplax.*

Table: *Host resistance to nematodes*[1]

Nematode species	*Crop*
	Highly effective
Ditylenchus dipsaci	Alfalfa, oat, red clover
Meloidogyne spp.	*Alfalfa, Prunus* (Nemaguard rootstock),*Lespedeza,* tomato, walnut (California Black)

Contd...

Nematode species	*Crop*
Heterodera avenae	Barley, oat
Tylenchulus semipenetrans	Citrus (*Poncirus trifoliata*)
Meloidogyne incognita	Cowpea, lima and common bean, soybean, tobacco
Globodera rostochiensis	Potato
Heterodera glycines	Soybean
Some developed resistance	
Heterodera avenae	Barley, oat
Tylenchulus semipenetrans	Citrus
Radopholus citrophilus	"
Meloidogyne incognita	Cotton
Rotylenchulus reniformis	"
Meloidogyne javanica	Cowpea, tobacco
Meloidogyne arenaria	"
Heterodera zeae	Maize[2]
Globodera rostochiensis	Potato
Globodera pallida	"
Meloidogyne spp.	"
Nacobbus aberrans	"
Pratylenchus vulnus	Prunus spp.
Xiphinema americanum	"
Criconemella xenoplax	"
Heterodera glycines	Soybean
Meloidogyne spp.	"
Rotylenchulus reniformis	"
Meloidogyne hapla	Tomato
Meloidogyne incognita	"
Meloidogyne javanica	"
Meloidogyne arenaria	"
No developed resistance	
Heterodera schachtii	Sugar beet, crucifers
Heterodera avenae	Wheat
Meloidogyne spp.	Barley, carrot, maize, cucurbits, eggplant, lettuce, groundnut, sugar beet, many others
Meloidogyne hapla	Most host crops
Rotylenchulus reniformis	Tomato, many host crops
Naccobus aberrans	Pepper, sugar beet, tomato

[1] *Largely after Roberts.*

[2] *Some sources identified (L.K. Krusberg, personal communication). no developed resistance for a number of important nematode-host combinations.*

The deployment of host-resistant cultivars has been accomplished largely on a regional basis. A world survey by Sasser and Freckman (1987) indicated that at least some nematode-host resistance or tolerance is available on one or more crops in all regions surveyed. In Africa, however, only four crops were reported to have tolerance or resistance: bean, cowpea, tomato and yam. The resistant host lists for other geographic regions is somewhat longer and South America had the most extensive list with 26 plant species. The list for Asia was intermediate with 18 crop plants exhibiting some nematode resistance.

The available host resistance to nematodes also has some limitations. For example, resistance *to Meloidogyne* spp. in bean and tomato plants is heat-sensitive, and thus may fail in hot production regions at lower latitudes. In addition, crop cultivars with resistance to *Meloidogyne* spp. are often susceptible to other associated nematode species. The latter limitation is not a serious restriction in fields that have a primary dominant or target nematode species. However, the restricted deployment of host-plant resistance in the large African continent reflects a critical regional and worldwide need for greater collaboration between nematologists and plant breeders in the development and deployment of this important tactic for reducing nematode problems.

Available host-plant resistance to nematodes should be deployed in carefully developed cropping systems. Monocultures with resistant cultivars frequently result in the appearance of biotypes of the target nematode that circumvent the resistance.

Supplemental chemical nematicides have come to the forefront in some regions as a means of suppressing the increase of nematodes on high-value crops such as groundnut or citrus. Repeated applications of low dosages through trickle irrigation may suppress nematode multiplication, resulting in the growth of a healthy crop.

Restriction of Current Crop Damage

Tolerant cultivars of suitable host species grow and yield well in spite of nematode parasitism. Although such hosts have near-normal shoot and root growth, nematode population densities build up to levels that often severely damage subsequent susceptible crops. For example, maize may support relatively high reproduction of M. *incognita* or M. *arenaria,* and still produce near-normal yields. Thus, such crops should be examined carefully before placing them in rotations with highly susceptible crops. A few plant breeders and nematologists are working to combine host-plant resistance and tolerance to nematodes. Resistant/tolerant cultivars clearly would have an advantage in that some resistant cultivars may be severely stunted by high population densities of the target nematode.

Integration of Strategies/tactics

As indicated earlier, Tyler (1933) suggested that two or more management tactics be combined to develop effective control programmes for root-knot nematodes, and also stated that the value and permanence of chemical or cultural treatment would be increased when followed by a wet fallow or resistant crop, with particular attention to the control of weed hosts. As highly efficacious, inexpensive chemical soil treatments have been removed from the market, the integration of available management tactics and the maximizing of the potential of cropping systems must receive much greater attention.

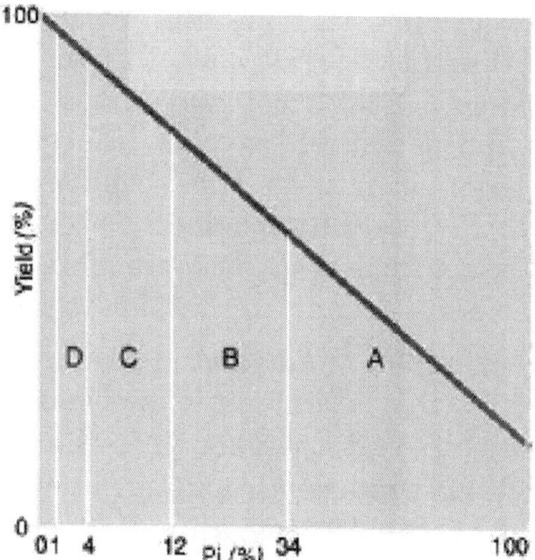

Figure: *Hypothetical damage function (relationship between yield and initial nematode population density, Pi), showing the percentage reduction in Pi and related increase in yield as four nematode control strategies (A-D) each with 66 percent efficacy are applied sequentially*

To date, most IPM nematology research has focused on population biology, host phenology, modelling, damage functions and overall pest management decision-making. Progress in these areas has clearly facilitated the development of better management programmes as well as enhancing our understanding of nematode-host interactions. Nevertheless, the integration and assessment of combinations of potentially compatible nematode management strategies and tactics have received only limited attention. This situation has developed, in large part, from heavy reliance on the use of nematicides, rotations where practical and resistant cultivars when available. The integration of host resistance and nematicides with other compatible tactics should minimize potential problems of loss of efficacy of given compounds or the appearance of nematode biotypes that attack resistant cultivars.

Integration of nematode management may be considered at two levels: within nematology; and across pathogen/pest disciplines. Within the context of nematodes, the application of multiple management tactics may be done over time or simultaneously, or both. The temporal approach focuses on season-to-season or year-to-year integration of tactics relevant to cropping cycles. This cycle of integration of tactics might focus on crop rotation with non-hosts, resistant cultivars, or both, with additional strategies such as nematicides, cultural practices or biological control being added as they become available. This approach could involve different control tactics on different crops for given years. The second approach described by Roberts (1993) for integrating nematode-control components, involved simultaneous use of two or more tactics. This approach requires compatibility among the tactics to be deployed.

Numerous examples of single and multiple pest/pathogen successes could be cited. Combinations of host resistance and tolerance, nematicides and rotations of resistant and susceptible potatoes with non-host rotation crops have provided a successful integrated management programme for the potato-cyst nematode. A rather different system focusing host resistance, crop rotation, residual host-root destruction immediately after harvest, and an appropriate nematicide where warranted, has proved to be highly successful for *Meloidogyne* spp. and certain fungi, bacteria, insects and weeds on tobacco in North Carolina.

Roberts (1993) also offered a theoretical but practical framework for the integration of nematode management tactics with only partial efficacy. This example depicts four hypothetical control tactics, A-D, with each having an efficacy of 66 percent. Thus, stepwise, residuals of the initial nematode population levels per successive tactic would be 34, 11.6, 2.9 and 1.3 percent, respectively. Information on the damage threshold for the target nematodes and the efficacy of the integrated management tactics deployed, would be essential to determining the number of the types of treatments necessary to achieve satisfactory crop yields. For highly aggressive nematode species such as *M. arenaria* on groundnut, an even greater number of these types of tactics might be required in a management programme. Clearly, more information on the potential interactions of available management tactics, whether antagonistic or synergistic, or simply additive effects, will be critical for the integration of nematode control strategies and tactics.

Conclusions and Outlook

Much progress has been made in characterizing nematode communities and in the assessment of some management tactics in the Near East. Research opportunities remain in these areas as well as in developing related damage functions/thresholds and the integration of compatible strategies/tactics. Candidates for this integration include rotation/cropping systems, resistant cultivars, fallow/destruction of residual infected roots, solarization, nematicides and soil amendments/biological control. Carefully developed research-extension priorities, and the use of some type of data/decision-support system, should facilitate the development and deployment of effective IPM programmes. Long-term IPM goals and priorities might focus on interdisciplinary IPM (all pathogen and pest groups) and sustainable crop-production systems with critical inputs from experts in technology development/transfer, institutional resources/training, and economic/social/cultural components.

Promising new technologies for nematode diagnosis and identification, critical to IPM, are on the horizon. In addition, novel management weapons, including transgenic plants with nematode resistance and/or biological control genes, new safer nematicides and more effective biological control agents should be forthcoming. Improved data manage-ment and decision-support systems should facilitate the integration of new and traditional IPM strategies and tactics.

Chapter 7

Nematode Biology and Management

Background Information

Plant-parasitic nematodes are nearly microscopic, worm-shaped animals virtually invisible to the naked eye when in the soil. They can cause significant plant damage ranging from negligible injury to total destruction of plant material. The severity of plant injury resulting from nematode activity depends on several factors such as the combination of plant and nematode species and prevailing environmental factors including rainfall, soil types, land contour, and culture practices.

Although a few nematode species feed on above ground plant parts, such as leaves, stems, flowers, and seeds, the majority of these parasites feed on underground parts of plants, including roots, bulbs, and tubers. Because of this below ground, "hidden" feeding activity, nematode damage to plants cannot always be diagnosed readily.

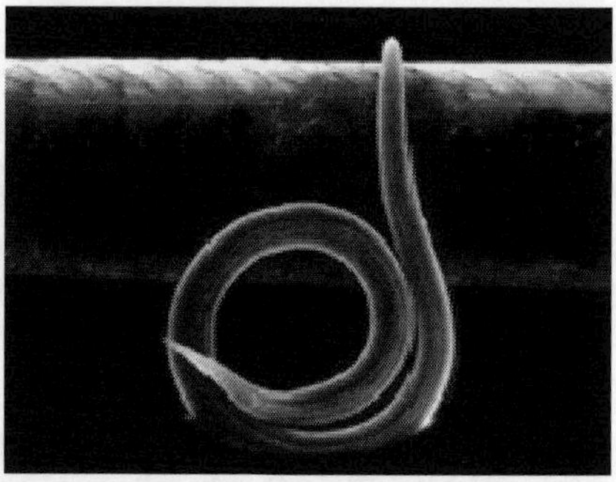

***Figure:** Scanning electron micrograph of a second-stage root-knot nematode juvenile propped up against a human hair (Eisenback)*

Meloidogyne incognita penetrating cucumber roots as seen with scanning electron microscopy. Many plant-parasitic nematodes have a wide host range while a few are host specific.

- Nematodes also feed and reproduce on many different weeds.
- The survival of some nematodes for at least one-year in moist soil without a host plant is common.
- Many kinds of plant-feeding nematodes can survive freely in the soil through winter-even if the soil is frozen.
- A few plant-parasitic nematode species can survive in the soil in a dormant state within cysts for many years.

Symptoms of Plant Damage due to Nematode Activity

- Initially symptoms are similar to lack of or improper fertilization, too little or too much water, improper soil pH, poor soil, or other environmental factors.
- Oval areas or irregular patches in the field with poor plant growth ranging from a few feet to hundreds of feet across.
- Poor, sickly growth, wilting, yellowing, stunting, and premature plant aging.

Field symptoms (patchy stunting and chlorosis) of soybean cyst nematode, heterodera glycines, on soybean.

Hundreds of different species of nematodes are associated with plants

- Not all nematodes are plant parasites.
- In many cases those that are plant parasitic occur in populations too small to cause serious plant injury.
- Extracting and identifying the nematodes from soil samples or infected plant parts are the only positive means for determining if and which nematode species are responsible for poor growth.

Detecting Nematodes in Soil and Plant Samples

To ensure proper collection and handling of samples submitted for nematode analysis, refer to the Nematode Diagnostic Laboratory "Sampling Procedures" fact sheet.

Figure: *Baermann funnels for extraction of motile nematodes.*

In addition, a fully completed "Sample Information" form must accompany each sample submitted for laboratory analysis. Both of these forms are available at county Maryland Cooperative Extension offices or this laboratory.

Nematode Management

- The individual situation determines the use of management tactics.
- The degree of plant damage varies with the kinds and numbers of nematodes present, the kind and variety of plant, and the cultural and environmental factors involved.
- Management tactics rarely destroy all plant-parasitic nematodes, but rather reduce the population(s) to levels below which economic damage occurs.
- When control measures are recommended, it is often necessary to repeat the measures periodically to maintain satisfactory plant growth.

The following widely recognized Integrated Pest Management (IPM) strategies are critical to management of plant-parasitic nematodes.

Prevention: This strategy should be the first line of defence. Prevention is the practice of keeping a population of nematodes from infesting a site and specific tactics include:

- Education regarding nematode presence, biology, and management.

- Survey sampling of an intended planting site to determine the degree of existing nematode activity.
- Use of nematode free transplants.
- Careful Examination of transplant roots before planting. Plants already infected when put into the ground will not grow well and can introduce nematodes to the site that attack present and future susceptible plantings.
- Weed control as a means of eliminating nematode access to alternate hosts.
- Sanitation practices such as cleaning of equipment between sites of operation.

Avoidance: practiced when nematode populations exist in a site, but their impact can be avoided or lessened through some cultural practice. Examples of avoidance tactics include:

- Sampling to determine nematode species and population levels.
- Choosing plant materials that are poor hosts.
- Practice crop rotations that include nonhost, resistant and susceptible crops when feasible.
- Employ cultural practices known to be optimal for plant maintenance.

Monitoring: includes tracking and proper identification of nematodes and includes:

- Soil and plant sampling for detection and trouble shooting.
- Routine sampling to maintain records of incidence and distribution. These samples form the basis for crop rotations, cultivar selection, economic thresholds, suppressive actions, and other management decisions.

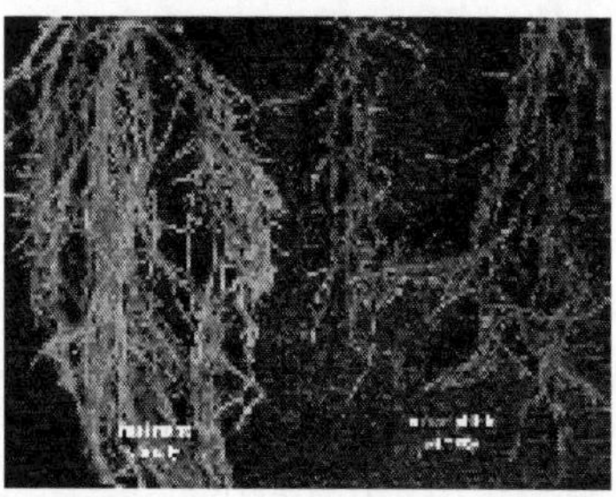

Figure: *Development of soybean cyst nematode, Heterodera glycines, on resistant (left) and susceptible (right) soybean varieties.*

Suppression: nematode suppressive tactics may become necessary to avoid economic crop loss where prevention and avoidance tactics are not successful. Suppressive tactics include:

- Addition of organic matter and green manure crops to encourage the release of microorganisms detrimental to nematode development or survival.

Chemical Management

- Chemical control involves the application of certain volatile (fumigants) and nonvolatile nematicides to the soil to reduce nematode populations. It is important here to note:
- There are some nematode problems for which there currently is no legal, effective nematicide.
- Chemical management reduces nematode populations for a limited period of time. Nematodes that escape treatment can resume feeding when the chemical is gone.
- Nematicides are severely restricted for landscape application.
- Because they are highly toxic, several nematicides that are registered for ornamental plants are restricted to commercial production.
- There is presently no effective nematicide that may be applied legally to ornamentals already planted in the landscape.

Plant Nematodes Attack All Plants grown in Florida. They cause farmers and nurserymen millions of dollars in crop loss annually, but also can cause problems in the urban world by damaging turfgrasses, ornamentals and home gardens. We are often unaware of losses caused by nematodes because much of the damage caused by them is so subtle that it goes unnoticed or is attributed to other causes.

Some scientists estimate that there are over 1 million kinds of nematodes, making them second only to the insects in numbers. However, few people are aware of nematodes or have seen any, because: Most nematodes are very small, even microscopic, and colourless; most live hidden in soil, under water, or in the plants or animals they parasitize; and relatively few have obvious direct effects on humans or their activities.

Of all of the nematodes known, about 50 percent are small animals living in marine environments, and 25 percent live in the soil or fresh water and feed on bacteria, fungi, other decomposer organisms, small invertebrates or organic matter. About 15 percent are parasites of animals, ranging from small insects and other invertebrates up to domestic and wild animals and man. Some of the parasites of animals are the largest nematodes known: some from grasshoppers can be several inches long, and one from whales can reach lengths of more than 20 feet! Only about 10 percent of known nematodes are parasites of plants.

Morphology and Anatomy. Plant nematodes are tiny worms usually 0.25 mm to 3 mm long (1/100″ to 1/8″) and cylindrical, tapering toward the head and tail. Females of a few species lose their worm shape as they mature, becoming pear-, lemon-or kidney-shaped. Plant parasitic nematodes possess all of the major organ systems of higher animals except respiratory and circulatory systems. The body is covered by a transparent cuticle, which bears surface marks helpful for identifying nematode species.

Life Cycle and Reproduction. The life cycle of a plant-parasitic nematode has six stages: egg, four juvenile stages and adult. Male and female nematodes occur in most species, but reproduction without males is common, and some species are hermaphroditic. Egg production by the individual completes the cycle. Most species produce between 50 and 500 eggs per female, depending on the nematode species and their environment, but some can produce more than 1,000 eggs. The length of the life cycle varies considerably, depending on nematode species, host plant, and the temperature of the habitat. During summer months when soil temperatures are 80 to 90°F, many plant nematodes complete their life cycle in about four weeks.

Nematode Feeding and Host-Parasite Relationships

Plant parasitic nematodes feed on living plant tissues, using an oral stylet, a spearing device somewhat like a hypodermic needle, to puncture host cells. Many, probably all, plant nematodes inject enzymes into a host cell before feeding to partially digest the cell contents before they are sucked into the gut. Most of the injury that nematodes cause plants is related in some way to the feeding process.

Ectoparasitic nematodes feed on plant tissues from outside the plant; endoparasitic nematodes feed inside the tissues. If the adult female moves freely through the 61 soil or plant tissues, the species is said to be Amigratory. Species in which the adult females become swollen and permanently immobile in one place in or on a root are termed Asedentary. Migratory endoparasitic and ectoparasitic nematodes generally deposit their eggs singly as they are produced, wherever the female happens to be in the soil or plant. Sedentary nematodes such as root-knot (*Meliodogyne* spp.), cyst (*Heterodera* spp.), reniform (*Rotylenchulus* spp.), and citrus (*Tylenchulus semipenetrans*) nematodes produce large numbers of eggs, which remain in their bodies or accumulate in masses attached to their bodies.

The feeding/living relationships that nematodes have with their hosts affect sampling methods and the success of management practices. Ectoparasitic nematodes, which never enter roots, may be recovered only from soil samples. Endoparasitic nematodes often are detected most easily in samples of the tissues in which they feed and live (burrowing and lesion nematodes), but some occur more commonly as migratory stages in the soil (root-knot and reinform nematodes).

Endoparasitic nematodes inside root tissues may be protected from those kinds of pesticides that do not penetrate into roots. Root tissues may also shield them from many microorganisms that attack nematodes in the soil. Ectoparasites are more exposed to pesticides and natural control agents in the soil.

Foliar nematodes (*Aphelenchoides* spp.) are migratory nematodes that feed on or inside the leaves and buds of ferns, strawberries, chrysanthemums and many other ornamentals. They cause distortion or death of buds, leaf distortion, or yellow to dark-brown lesions between major veins of leaves. Other nematodes that attack plants above ground, but are not common in Florida, cause leaf or seed galls. Still others cause deterioration of the bulbs and necks of onions and their relatives.

Chapter 8

Plant Parasitic Nematodes

Nematodes are microscopic eel-like organisms that live in soil and water. Nematodes are the most abundant multicellular organisms on earth. Most soil dwelling nematodes are beneficial organisms that play a role in the break down and release of nutrients from organic matter. Some beneficial nematodes prey on other nematodes as well as soilborne insect, fungal and bacteria pests. Unfortunately there are several species of nematodes that feed on or in roots, stems or bulbs resulting in significant yield reduction in both field and horticulture crops grown in Ontario.

Plant parasitic nematodes possess a hollow stylet, mouth part which is like a hypodermic syringe. The stylet is forced into plant cells and enzymes are injected to decompose the cell content. The nematode withdraws the partially digested cell contents through the stylet. Some nematodes such as the root knot and the cyst nematodes establish a specialized feeding site where they remain for the rest of their life cycle. Other nematodes such as the root lesion nematode burrow into the root, feeding and causing damage as they move through the root. When infested plants and roots die in the autumn, root lesion nematodes will move out of the roots into the soil.

Plant Parasitic Nematodes in Ontario

Several different types of plant parasitic nematodes inhabit different regions of Ontario. These include the soybean cyst nematode (*Heterodera glycines*), which is spreading throughout parts of Southwestern Ontario, the oat cyst nematode (*H. avenae*), the sugar beet cyst nematode (*H. schachtii*), the northern rootknot nematode (*Meloidogyne hapla*), bulb and stem nematode (*Ditylenchus dipsaci*), dagger nematode (*Xyphinema sp.*) and the root lesion nematode (*Pratylenchus penetrans*). Most cyst nematodes tend to have limited host range, except for the sugar beet cyst nematode that can infect over 220 hosts including most cole crops. The bulb and stem nematode has a wide host range and several races. In Ontario it tends to only cause significant damage to onions and garlic. The Northern rootknot nematode also has a wide host range and is particularly damaging to carrots. The root lesion nematode is considered the most economically important plant parasitic nematode in Ontario fruit and vegetable production. Root lesion nematodes have a wide host range including many native weeds.

Nematode Life Cycle

Most plant parasitic nematodes lay eggs in the soil or roots of host plants or are retained within the female body or cyst. After the eggs hatch, the juvenile nematodes swim to other nearby plant roots and feed on them. Damage caused by nematodes in many crops can also provide an infection site for other disease causing organisms, which further reduces yields. Nematodes complete their life cycle within three to six weeks during the growing season depending upon available moisture and temperature. Extreme moisture and temperatures will kill some species of nematodes.

Symptoms of Nematode Damage

Symptoms caused by nematode infestation differ depending on the crop and the type of nematode pest. Nematode damaged plants usually occur in patches or along a row. Infested plants may appear stunted, wilted and unthrifty. Nematode feeding also causes symptoms such as yellowing, stem twisting, crown and bulb bloating, root galls and root forking and distortion. High soil population levels of plant parasitic nematodes can cause death of young plants. Frequently the damage caused by nematodes goes undetected because of the difficulty of diagnosing nematode damage and infestation, resulting in the loss of millions of dollars in crop production annually.

Taking soil and/or root samples at the proper time of year for analysis of nematode populations by a qualified laboratory accredited to identify and enumerate nematodes, can help growers avoid or reduce the potential of nematode problems in their crops.

Root lesion nematodes cause small scratch-like lesions on feeder roots providing an avenue for other root rotting organism to infect. In fact root lesion nematodes have frequently been associated with disease complexes involving other soilborne pathogens resulting in significantly greater disease and reduced yields compared to either the fungal pathogen or nematode infection alone. Some parasitic nematodes such as the dagger nematode are vectors for plant viruses.

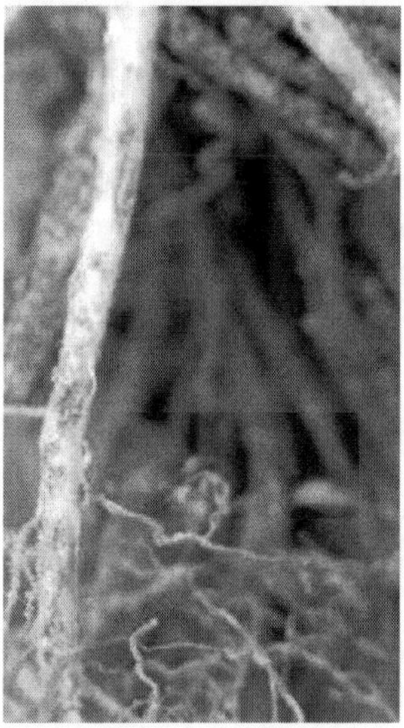

Figure: *Scratch-like lesions on strawberry root caused by root lesion nematodes.*

Roots that have small pearl-like structures along feeder roots may be an indication of cyst nematode. The small cysts may be white, yellow or brown in appearance depending on the species of the cyst nematode. Northern rootknot nematodes cause roots to appear stubby or swollen and often result in excessive secondary root branching giving the root a hairy appearance. Growers who notice patches of plants with unusual root symptoms, severe root rot or a patch that appears to decline quicker than the rest of the field may suspect a nematode problem and should consider sampling roots and soil for nematode analysis.

Figure: *Root swelling or knot symptoms on strawberry roots caused by rootknot nematodes.*

When to Take Root Samples

Root and soil samples containing roots can be taken at any time as long as the soil is not frozen. During the active growing season, however, nematodes live and feed inside or along roots particularly during hot dry seasons.

If nematodes are suspected of contributing to the decline of a particular area of a young crop during the growing season, collect entire root systems with surrounding soil separately from plants with symptoms and plants without symptoms. If the decline is noticed in a fruit tree orchard, vineyard or other perennial crop, carefully dig and sample from the feeder root zone approximately 10-20 g fresh weight of roots from the infected plants and submit for analysis. Do not sample the roots from dead plants because the nematodes will have already died or moved away from dead roots into the soil. Place samples in a plastic bag out of direct sunlight and in a cool place during transportation to the diagnostic lab.

When to Take Soil Samples

The best time to sample soil for nematode population assessment is in the spring after the soil has warmed up or during the fall, soon after harvest. Do not take nematode samples when fields are very wet. Fields with a history of nematode problems may be sampled routinely to determine if the nematode population is approaching or has exceeded an economic threshold.

Soil populations of most plant parasitic nematodes tend to be highest in September and October after crops have senesced and died. This is the best time of year to sample for nematodes. Sampling in the early fall allows growers time to make decisions on whether to fumigate during the fall or spring or what crop should be planted the following spring. It also allows time to implement an integrated management strategy prior to growing a susceptible crop in that field. Sampling in the spring prior to planting a crop may also be reliable.

Where to Soil Sample

Where to sample soil for nematode asessment depends on the purpose for taking the soil sample, the type of crop in the field, and the type of nematodes being sampled. If the purpose of sampling soil for nematodes is to diagnose a problem during the growing season in a row crop, take eight to 10 soil cores from areas where plants are unhealthy or near plants along the margin of a severely affected area. Sample another eight to 10 soil cores separately from areas of healthy growing plants for comparison. When sampling soil from row crops during the growing season, or from trees or perennial crops, it is very important to get the feeder roots of the crop in the soil sample, since this is where many nematodes live.

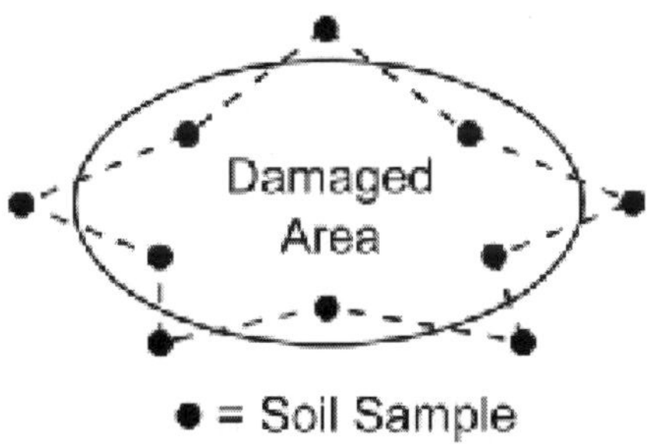

Figure: *Sampling pattern for damaged area or infected patch in a crop.*

For individual fruit trees or ornamental shrubs suspected of being infested with nematodes, it is best to take soil samples from just below the drip line and in the area between the outer branch tips and the tree trunk.

If the purpose of sampling a field is to determine whether the nematode population has reached an economic threshold in a row crop, take soil cores within the row of actively growing plants to obtain samples that contain feeder roots. When sampling from fallow fields, in the autumn after the crop has senesced or in the spring prior to planting, it is best to walk in a Z, W or M pattern across the field. The soil sample should represent no more than 2.5 ha.

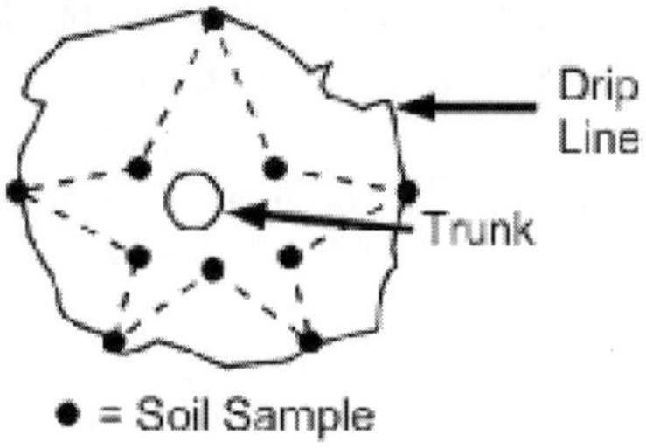

Figure: *Sampling pattern for individual tree or shrub.*

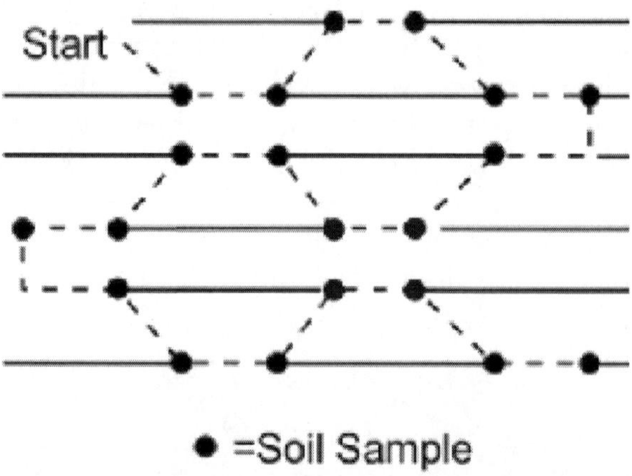

Figure: *Soil sampling pattern for row crops.*

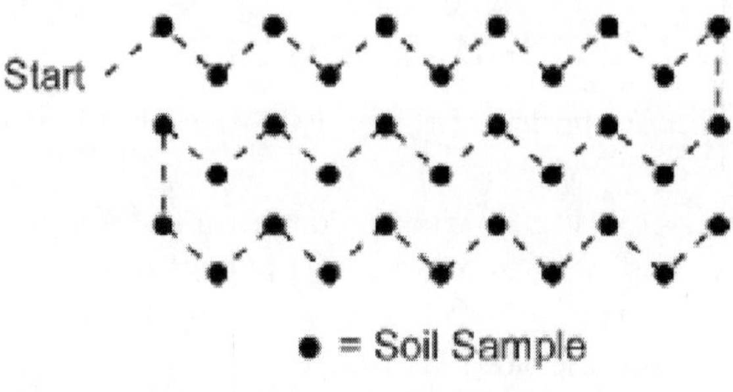

Figure: *Sampling pattern in the autumn after the crop has senesced or in the spring prior to planting. This pattern can also be used for fallow fields and crops that are not planted in rows.*

How to Sample Soil

Nematodes are rarely distributed evenly throughout a field and nematode populations fluctuate throughout the growing season. Soil should be sampled approximately 20 cm (8 in.) deep using a 2.5 cm(1-inch)-diameter soil core probe. Alternatively, soil can be sampled with a narrow bladed shovel or trowel; however, this method is less reliable than using a soil core probe.

Extremely wet, dry, hot or cool seasons can influence the population levels particularly in the top 2.5-5 cm (1-2 in.) of soil. Discard the top 2.5-5 cm (1-2 in.) of soil where nematodes would not usually live due to extreme environmental conditions.

Collect soil cores in a clean bucket, mix the soil thoroughly but gently and place in a labelled plastic bag or container. Never allow soil samples to heat up or dry out. Place soil samples in a cooler with ice until they can be stored in a fridge or analysed for nematode populations.

Figure: *Sampling soil with a 2.5 cm- diameter soil probe*

Number of Soil Cores/Area

The number of soil core samples required to estimate nematode soil population levels, depends on the size of the area under investigation. The sample submitted to the laboratory should not represent more than 2.5 ha.

Enough soil to give a good representation of the soil population is all that is necessary. The chart below is a guide of how many cores are necessary to make up a representative sample.

If soil type changes within the field, take separate samples from each soil type. Send the soil samples to a pest diagnostic clinic or laboratory that is qualified to isolate, identify and enumerate nematodes.

***Table:** Number of soil core samples/area required to estimate nematode populations*

Area	*Number of soil cores/sample*
< 500 m^2	8- 10
500 m^2 - 0.5 ha	25 - 35
0.5 ha - 2.5 ha	50 - 60

Economic Thresholds

Economic thresholds are based on initial soil population levels that will multiply over the growing season and cause economic damage to the crop. The economic threshold is expressed as the number of nematodes in a kilogram of soil and is often different for each crop and each nematode species. Threshold levels from as high as 1,000 root lesion nematodes/ kg soil for most vegetables to as few as 500/kg of soil in strawberry can significantly reduce yields. If the nematode soil analysis report indicates populations higher than the threshold, an integrated nematode management strategy should be implemented.

Root Diffusates and Plant Parasitic Nematodes

The effects of root diffusates on plant parasitic nematodes are discussed. Factors in root diffusates can stimulate or inhibit the hatching of the juveniles, affect their molting, or attract the nematodes to their hosts.

Root diffusates can affect different stages of plant parasitic nematodes by stimulating or inhibiting the hatching, activating the juveniles to molt, and attracting the nematodes to the host. Although much work have been done several questions still remain to be answered.

The nature of different nematodeattractants in diffusates are poorly known as weil as the mode of action, and the response of the nematodes. Most of the papers published so far concern the hatching factors of the cyst forming nematodes (*Heteroderidae*), with some notes on root-gall nematodes (*Meloidogyne* spp), and a few other genera.

Diffusates

The hatching factor in root diffusates has been extensively studied for the potato cyst nematodes and for the soybean cyst nematode (*Heterodera glycines*). The hatching factor for the potato cyst nematode has not, so far, been isolated. The active chemical seems to be rather unstable in soil. It is apparently acidic in nature, stable at pH 1-7, water soluble, nonvolatile, containing a lactone group, and one or two hydroxyl groups.

It is believed to have an empirical formula of approximately $C_{18}H_{24}O_8$. A hatching agent for the soybean cyst nematode, glycinoeclepin A, which is a diabasic acid with a molecular weight of 446, has recently been isolated from the roots of *Phaseolus vulgaris*. The hatching agent in the soybean roots remains, however, to be isolated.

Hatching

Most nematode species hatch without delay once the juvenile has reached a certain stage of development in suitable environmental conditions. Several cyst nematode species, however, need a stimulus released into soil by host plants before hatching is activated.

Some *Heterodera*-species, such as *H. glycines* and *H. goettingiana* hatch in root diffusates from host plants, while species like *H. avenae* and *H. schachtii* are able to hatch both in host-root diffusates and in water.

In the genus *Globodera* most species have a narrow host-range, and their hatching usually depends on exposure to host-root diffusate. The most important species in this genus are the potato cyst nematodes with only very few hosts in the family *Solanaceae*.

Only about 2% of the juveniles hatch in water the majority of the juveniles beeing activated only after exposure to root diffusates from potato, tomato, an a few other hosts. The spontaneous hatching of the potato cyst nematodes frequently reported in fallow conditions is assumed to be caused by substances produced by other soil organisms.

Some nonhosts, such as some grasses and cereals have been observed to stimulate the hatching of the potato cyst nematodes, although the hatching seems to be slower in non-host root diffusates as compared to the potato root diffusate.

Indications of synergistic effect of root diffusates on hatching of the potato cyst nematode have also been reported. More juveniles hatched in vitro when potato and nonhost root diffusates were present together as compared with potato root diffusate alone.

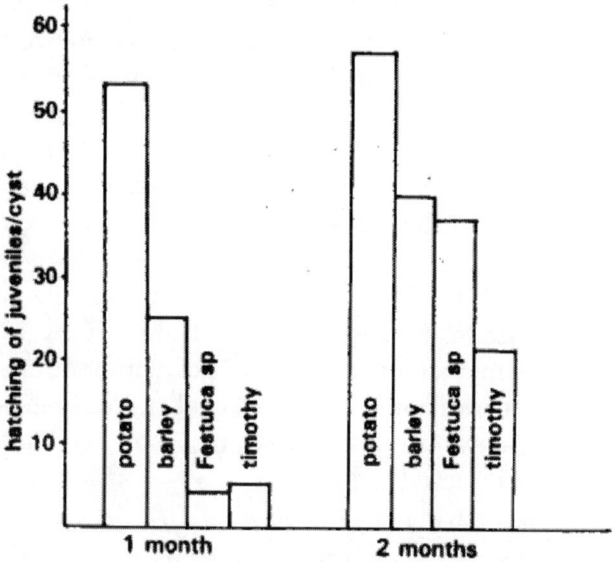

Figure: *Total numbers of 2nd stage juveniles of* G. rostochiensis *hatching in vitro in root diffusates from potato and some non-hosts after 1 and 2 months respectively. Means of four replicates.*

Nonhosts can also have an inhibitory effect on cyst nematode hatching. The hatching of the potato cyst nematodes is inhibited by root diffusates from mustard and some *Umbelliferae*. Root diffusates of *Tagetes* spp., however, did not inhibit hatching of the potato cyst nematode despite the naturally occurring nematicide in the roots. The mode of action of the hatching factor(s) in the root diffusates is not completely understood.

There are observations of an increase in oxygen consumption of unhatched potato cyst nematode juveniles within 24 hours after exposure to potato root diffusate, and changes in adenylate energy charge of the cyst contents, which led to altered metabolism within the cysts changing presumably the permeability of egg shells. The hatching factors in potato root diffusate also activate the unhatched juveniles to move vigorously, and finally to cut a straight slit across the egg with the stylet.

Molting

The effect of root diffusates on molting of plant parasitic nematodes has only been reported of the ectoparasitic genus *Paratylenchus. P. projectus* and *P. dianthus* juveniles survive for long periods in soil without hosts. Very few larvae molt in water, while the majority of them molt in carnation root diffusate. Similar behaviour has been reported of *P. nanus*.

Host Findning

The movement of plant parasitic nematodes to their hosts frequently involves a random component, but chemoreceptors are probably important at least when the nematode is close to the host roots. Gradients of CO_2 concentrations seem often to be a main attractant to some species, eg. *Pratylenchus penetrans* and *Meloidogyne hapla*, but some species, for instance *H. schachtii* and *P. scribneri* fall to show a marked aggregation around roots of nonhosts, and they may be influenced by specific stimuli present near the host roots.

The response of the nematodes to the attractants from the roots is greatly influenced by environmental factors such as moisture and aeration, but also by the developmental stage of the nematode.

Synthesis

The stimulating factors in root diffusates play an important role for many plant parasitic nematodes, especially for the species/genera highly specialized to live as plant parasites. The majority of the studies consider cyst nematodes, not only because of the noted efficiency of root diffusates, but also because of the economic significance of the cyst nematodes.

Isolating of at hatching factor for any economically important cyst nematode species may prove of great value if this were to result in the synthesis of a cheap analogue chemical useful for control.

Root Lesion Nematodes

Root lesion nematodes are minute, destructive pests of plant roots. They are wormlike and to the naked eye look like tiny threads of cotton. They can only be seen in detail under the microscope. They live in the soil, moving in the free water between soil particles; they enter the plant roots by pushing their head and using their stylet (which is like a jackhammer in the mouth part) against the cells of the roots.

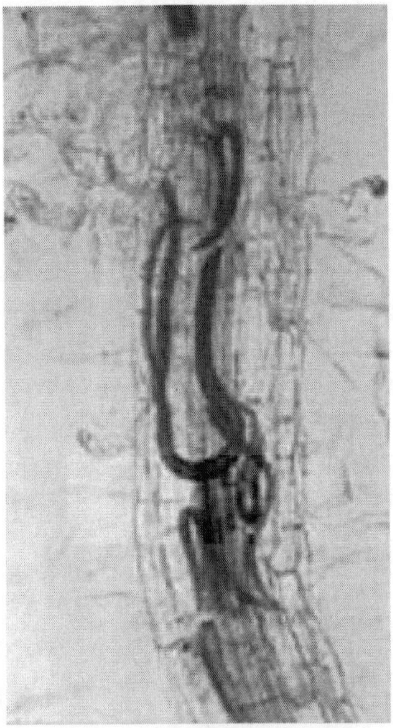

Figure: *Root Lesion Nematodes Inside a Wheat Root*

Once inside the roots, the nematodes move through the plant until they reach a feeding site. They then feed on the contents of the plant cells and lay eggs, causing damage to the plant roots and root malfunction. Dark lesions may develop on the roots but, more commonly, we see non-specific symptoms in the plant tops, such as lower leaf yellowing, stunting and poor tillering, resulting in unhealthy crops that do not yield well.

Two Species of RLN in North

There are two species of root lesion nematodes found in the northern grains region-*Pratylenchus thornei* and

Pratylenchus neglectus, with *P. thornei* most common in this region. It is important to identify these nematodes to species level because they each attack different crops and will even favour different varieties within crop species-wheat is a good example of this. Thus management for each of these nematodes is quite specific. The Department of Primary Industries and Fisheries (DPI&F) can test farms for nematodes.

On the left, a susceptible wheat had been planted in year 1, followed by a susceptible mungbean. On the right, resistant canary seed in year 1 was followed by a resistant sorghum, giving twice the yield of Strzelecki wheat in year 4.

Host Crops

Wheat is the preferred host of root lesion nematodes, but other crops such as chickpea can also suffer yield loss. These susceptible crops allow nematode populations to multiply; resistant crops do not. We have shown that many other crops are good hosts of the root lesion nematodes but may be tolerant-the plants continue to grow and yield well, but nematode numbers increase. For example, mungbeans will leave high populations of *P. thornei* in the soil, but suffer little or no yield loss; and in paddocks infected with *P. neglectus*, sorghum will allow populations to build up. So it is easy to be unaware that nematode populations are slowly increasing in your soil until a susceptible and intolerant crop, such as wheat or chickpea, is planted. The good news is that we have shown that wheat yields can be doubled by careful selection of rotation crops.

Management

The key to managing root lesion nematodes depends on three important points:

- accurately identifying the nematodes
- farm hygiene to prevent the spread of infected soil on machinery
- growing tolerant and resistant crops to keep root lesion nematode populations at manageable, low levels.

The tolerance and resistance of wheat varieties to both nematode species is published every year in the DPI&F publication 'Wheat varieties for Queensland'. Some resistant rotation crops, such as millets and panicums, appear to actively lower *P. thornei* populations even more that leaving the soil in bare-fallow.

Tillage, fallow and chemicals don´t work.

Bare-fallow or tillage do not reliably control root lesion nematode populations because the nematodes have developed clever survival mechanisms such as moving deep into the soil profile.

In the northern grain region, root lesion nematodes typically have peak populations at 60 cm depth and nematodes can be present even at 1 m depth

During drought or fallow periods, the nematodes dry out, in a state called anhydrobiosis (meaning 'life without water'), and wait until they receive a signal from plant roots and soil water to become active and attack again. Chemical control or fumigation is not an option because of high costs, and again the survival of the nematodes deep in the soil profile means that chemicals rarely penetrate the soil deep enough to control the nematodes.

Take-home messages.

- Have your paddocks checked for nematodes and know which species are present.
- Employ farm hygiene measures to limit the spread of the nematodes.
- Grow tolerant crops for maximal yields and rotate with resistant crops to keep nematode numbers at low, manageable levels.

Chapter 9

Biological Control of Nematodes: Prospects and Opportunities

In the past 20 years three developments have occurred which have had significant effects on the prospects and opportunities for the biological control of plant-parasitic nematodes. First, several nematicides have been withdrawn from the market because of health and environmental problems associated with their production and use. As a result of this, and increasing public concern over the use of pesticides in food production, there has been increased interest in the development of alternative methods of control, including the use of biological agents. Second, it has been demonstrated in several soils that nematophagous fungi and bacteria increase under some perennial crops, and under those grown in monocultures, and so may control some nematode pests, including cyst and root-knot nematodes. Such nematode-suppressive soils have been reported from around the world and include some of the best documented cases of effective biological control of nematode pests. Finally, a number of commercial products based on nematophagous fungi and bacteria have been developed, but all so far have had only limited success. Their use has been based on empirical research, and it is instructive to consider what might be the key factors for a successful biological control agent for nematodes in order to identify the reasons for the general failure of the products that have been developed.

Biological control is more inconsistent, less effective and slower acting than control normally achieved with chemicals. Although improvements in performance might be expected from more research on individual agents, it seems likely that these limitations are inherent in most biological control agents and that their successful application will depend on integration with other control measures.

The most studied example of natural control of a plant-parasitic nematode concerns the decline of populations of the cereal-cyst nematode, *Heterodera avenae,* under monocultures of susceptible cereals in many soils throughout northern Europe. This is an example of an induced suppression in which the nematode increases to damaging population densities in the second and third cereal crops, but usually declines thereafter to infestations of <5 eggs/g soil, which cause little loss of yield in northern European conditions; it is essential that the nematode is abundant in the early years of the monoculture to support the build-up of the microbial parasites.

The decline in nematode populations is mainly caused by two parasitic fungi, *Nematophthora gynophila* and *Verticillium chlamydosporium,* which attack the developing female on the root surface; in suppressive soils 95 to 97 percent of the females and eggs are destroyed. Thus the natural control of cereal-cyst nematode in a range of soils is predictable and effective, but slow acting. Research on the manipulation of natural control has been limited and attention has concentrated on the introduction of specific agents to provide more rapid control that might be commercially exploitable.

It has proved difficult to develop a biological control agent that is effective worldwide for any soilborne disease. Despite much research effort only two agents have had widespread success: *Phlebia gigantea* for the control of *Heterobasidion annosum,* which spreads from tree stumps to the roots of adjacent trees, *and Agrobacterium radiobacter* that is applied as a root dip to transplants for control of crown gall caused by *A. tumefasciens.* For both, the agents are applied in high concentrations as inundative treatments to a readily accessible site of action (the cut surface of a tree stump or the bare roots of a transplant) and protection from the diseases for a relatively short period provides long-term control. In most situations where nematode control is required, there is a need to provide long-term protection to a relatively inaccessi-ble and growing root system without the use of inundative treatments that are likely to be impractical and uneconomic. Hence the development of biological control agents for plant-parasitic nematodes is likely to be difficult and to require a detailed understanding of the biology and ecology of the agent and the nematode target.

The farming system in which biological control is applied has a marked effect on the way the agent is used. In general, growers in developed agriculture require an agent that can be applied to crops grown in monocultures over large areas using standard application machinery, so a formulated product with a good shelf-life that can be applied at low dosages is required; seed treatments are preferred for most arable crops. Little research has been done on the mass production and formulation of biological control agents for nematodes. Some organisms, such as rhizosphere bacteria, can be applied as seed treatments (Oostendorp and Sikora, 1989), but such applications tend to provide short-term control and are only useful in reducing the invasion of roots by nematodes that have a single generation in the growing season. In subsistence farming systems, crops tend to be grown in mixed stands in relatively small areas (often less than one hectare) and labour inputs are often large. As a consequence, relatively large application rates (up to one tonne per hectare) of an unformulated agent could be mixed into the soil by hand, as long as the organism could be produced cheaply and locally. Thus, the initial exploitation of biological control agents for nematodes may be in developing countries. However, if agents are only effective against specific nematode pests, and their efficacy is dependent on pest densities, then their effective use will require expert advice that may not be available in many developing countries.

Exploitation of Natural Enemies

The biology, ecology and potential of biological control agents for nematodes have been extensively reviewed in recent years. Nematologists have identified natural enemies with a range of modes of action similar to those currently studied by plant pathologists for the control of soilborne diseases. Comments on the advantages and limitations of the major groups of organisms with potential as biological control agents are summarized. It must be stressed that several organisms that are effective natural enemies of nematodes in the field, may have limited potential as biological control agents for application by growers. For example *Nematophthora gynophila* is a causal agent of cereal-cyst nematode decline in many soils, but its limited host range, complex requirements for *in vitro* culture, and need for soil moisture levels to be at field capacity to ensure infection, mean that its potential use is too limited to warrant further development as a biological control agent.

The stage in the nematode life cycle attacked by the control agent has a profound effect on the damage to the crop and the level of population control. Therefore, trapping fungi and rhizosphere bacteria that attack the second-stage juveniles of cyst and root-knot nematodes may significantly improve crop growth but is unlikely to prevent nematode populations increasing, especially those species that have more than one generation in a growing season. In contrast, those parasites that attack developing females and eggs act like a partially resistant cultivar in that initial nematode invasion and plant damage is not prevented but multiplication of the nematode is significantly reduced. Root colonizing fungi such as the mycorrhizae and endophytic species such as *Fusarium* spp. may reduce both nematode invasion and development.

Sikora (1992) reviewed a range of control measures, including crop rotation, partial soil sterilization, soil amendments and nematicides, that could be combined to increase the activity of naturally occurring biological control agents. Such measures could also be used to improve the performance of agents added to soil. Partial soil sterilization by methods such as solarization reduces nematode infestations and also reduces the competition from the residual soil microflora, enabling the biological control agent to establish more readily. Soil amendments may also reduce nematode infestations and, by providing an energy source, help to increase numbers of facultative parasites; pre-colonized substrates are most effective in establishing nematophagous fungi in soil. Soils that are naturally suppressive to some nematodes may be used to shorten rotations of susceptible crops and improve the performance of nematicides and resistant cultivars. The use of crop cultivars which are tolerant of nematode attack is likely to be very important for the successful deployment of biological control agents, that provide less effective nematode control than most nematicides. Agents such as rhizosphere bacteria and rhizosphere competent fungi depend on root exudates for their proliferation on roots; exudation from roots differs markedly between plant species and cultivar and affects the efficacy of these agents.

Environmental Concerns

Soil conditions, nematode species, rate of development and density and host plant all have a considerable effect on the establishment and activity of biological control agents for nematodes. It is essential to understand these interactions for effective control.

In temperate zones soil temperatures range between 10° and 15°C for much of the growing season whereas in warmer climates temperatures of 20° to 25 °C are normal at 10 cm depth. Too often, research workers in temperate climates have tested the effects of biological control agents in pot experiments in greenhouses at temperatures considerably above those which occur in the field; data from such tests are of limited value. Temperature has a direct effect on the growth and sporulation of the agent and on the rate of development of the nematode target. Moisture in soil is rarely limiting for the growth of most fungi but affects spore dispersal, especially of zoospores. Bacteria are more sensitive to soil moisture levels than fungi, but are unlikely to be affected by levels that enable nematodes to remain active. Soil texture and structure influence nematode activity and the growth and spread of microorganisms. The great bulk of soil (2 500 tonnes per hectare) to plough depth (20 cm) makes the thorough incorporation of biological control agents difficult and broadcast treatments are unlikely to be economic. The use of agents in arable crops is likely to depend on those organisms that proliferate in the rhizosphere from in-row applications or from seed treatments. The residual soil microflora competes with the introduced agent for scarce energy sources and can significantly affect the performance of the agent even when it is added to soil in a pre-colonized substrate.

Most plant-parasitic nematodes remain mobile throughout their life cycle and microbial parasites must produce traps or adhesive spores to infect them. Nematodes with sedentary stages such as cyst and root-knot nematodes may be parasitized from vegetative hyphae in the rhizosphere without the production of specialized infective structures apart from an appressorium. Fungi such as *Verticillium chlamydosporium* infect female cyst nematodes that are exposed on the root surface for several weeks during their maturation; depending on the time of infection, the fungus has a range of effects including a reduction in the number of mature females and their fecundity, and parasitism of the eggs. Usually only the eggs and not the females of root-knot nematodes are exposed in the rhizosphere, and these complete their embryonic development and hatch within about ten days at 25 °C. Hence, egg-parasitic fungi have less time to kill root-knot nematode eggs compared with those of cyst nematodes. Non-parasitic microorganisms that degrade soil amendments and release nematicidal compounds, such as the bacterium which degrades chitin to produce ammonia, are likely to kill most nematodes in soil. Similarly, many endophytic fungi may compete with a wide range of nematodes in roots although those which are sedentary are likely to be more susceptible.

The host plant influences the rate of nematode development and fecundity and the establishment of some natural enemies in the rhizosphere. Despite these effects the role of the host plant has rarely been considered in research on the biological control of nematodes.

Development of a Biological Control Agent for Nematodes: An Ecological Approach

Too often, biological control agents have failed because they have been used before a basic knowledge of their ecology and biology has been established. The importance of such knowledge can be seen from work at Rothamsted aimed at the development of *V. chlamydosporium* as a biological control agent for root-knot nematodes. Stirling (1991) describes the design and methods used in the conduct of biological control experiments, which are more complex than those required to assess the efficacy of a nematicide. He identified five key aspects in setting up an experiment to evaluate a biological control agent and these are presented below in a slightly modified form.

- The test organism and any organic amendment should be applied at practical application rates; 0.1 percent w/w soil is equivalent to 2.5 tonnes/ha and should represent a maximum dose. Tests should always be performed in a non-sterilized soil with a natural residual soil microflora.

- Appropriate treatments as well as an untreated control should be included if the organism is added with a substrate. These treatments should include the substrate alone, the organism alone, and the autoclaved colonized substrate. Too often, untreated controls are compared only with large applications of the organism and substrate and this does not allow separation of the effects of the agent from the effects of the substrate. In several tests reported in the literature, application of the substrate alone has decreased nematode populations to the same extent as the substrate colonized by the agent, and there is no clear evidence of biological control.
- Population densities of the agent under test should be monitored to ensure that it has survived in soil throughout the period that activity against the nematode target is required. Such monitoring may require the development of selective media, which can be a difficult and time-consuming task.
- Nematode mortality caused by the organism under test should be measured to assess whether differences between nematode population densities in treated and untreated soil relate to the levels of kill caused by the agent. Infection levels are relatively straightforward to estimate for most parasites, but repeated sampling is required to determine total kills. The effects of agents which produce toxins or have indirect effects on nematodes through competition, the modification of root exudates or the colonization of feeding cells, can only be measured by assessing their impact on nematode development.
- The impact of the soil environment, host plant and nematode should be tested as these are likely to affect the efficacy of the biological control agent, and could account for the lack of activity of potential agents in specific test conditions.

In a review of the literature less than 15 percent of experiments purporting to demonstrate biological control caused by *Paecilomyces lilacinus* satisfied the above criteria. Although this is an unsatisfactory situation that must be remedied, the difficulties in conducting carefully controlled and monitored experiments should not be underestimated.

Isolates of *V. chlamydosporium* differ markedly in their growth and sporulation *in vitro*, and in their virulence, saprophytic competitiveness and rhizosphere competence. Such differences between isolates of the same species of microorganism are common and there is a need for simple laboratory-based screening methods to select the most promising isolates for further testing. *Verticillium chlamydosporium* isolates were collected from infested nematode females and eggs in suppressive soils around the world. Hence, they were isolated from the niche in which they would be required to be active as a biological control agent and from soils most likely to yield active isolates. Once isolated in pure culture using standard techniques, the *in vitro* growth requirements were determined.

Virulent isolates were selected by counting the number of nematode eggs parasitized after exposure to the fungus on agar in a standard test. Tests for proliferation in the rhizosphere and growth in soil were also used to select promising isolates. Although simple laboratory-based screens help eliminate many isolates that show insufficient activity to justify further testing, selected isolates will not necessarily be active in the field. As it may take 10 to 16 weeks to investigate adequately the performance of different isolates against cyst and root-knot nematodes in pot tests, relatively few can be screened.

The method of mass culturing of *V. chlamydosporium* for experiments can have a marked effect on the subsequent survival and proliferation of the fungus in soil. Inoculum produced in shaken liquid cultures consists mostly of hyphae and conidia, which require an energy source to ensure proliferation in soil, whereas on solid media large numbers of chlamydospores are produced and these can be added to soil in aqueous suspension and rapidly establish the fungus.

The development of a semi-selective medium has enabled detailed studies to be made on changes in relative abundance of the fungus in soil and on roots. Some isolates of *V. chlamydosporium* may be extremely abundant in soil but unless they are capable of colonizing the rhizosphere they do not parasitize the eggs of root-knot nematodes. Growth in the rhizosphere differs markedly between plant species, e.g. tomato, cabbage and maize roots support much growth, whereas sorghum, pepper and cotton are poor hosts. Colonization by the fungus is confined to the rhizosphere and rhizoplane and there is no spread into root tissue; no lesions have been observed on roots grown in soil treated with *V. chlamydosporium* and there have been no detrimental effects on the growth of a range of crop species. A tenfold reduction (from 10^4 to 10^3 chlamydospores/g soil) in the amount of fungus applied to soil had no effect on the extent of colonization in the rhizosphere; the ability to proliferate on the root surface where the fungus is required to control nematodes is an important characteristic which may allow significant reductions in the amount of inoculum applied to soil.

The efficacy of *V. chlamydosporium* as a biological control agent for root-knot nematodes is affected by three key factors: the amount of fungus in the rhizosphere; the rate of development of eggs in the egg masses; and the size of the galls in which the female nematodes develop. In large galls female root-knot nematodes may produce egg masses which remain within the gall and are not exposed to parasitism by *V. chlamydosporium,* which is confined to the rhizosphere. Hence, *V. chlamydosporium* is less effective in controlling root-knot nematodes in heavily infested soils and on highly susceptible crops because large galls are formed on the roots and many eggs escape parasitism.

Verticillium chlamydosporium is unlikely to be useful in these situations where a grower would normally apply a nematicide. Also, at temperatures above 25°C eggs may complete their embryonic development and hatch before the fungus has completely colonized the egg mass; at 30°C about 30 percent of eggs of three root-knot species hatched and the second-stage juveniles escaped from the egg mass before the eggs were killed. These studies in pot tests, if supported by field experiments, help to define the conditions in which *V. chlamydosporium* might be used successfully for control of root-knot nematodes. It is only from such detailed studies that the limitations and requirements of the fungus can be assessed.

Verticillium chlamydosporium is not a replacement for nematicides but, despite its limitations, it may be a useful management tool. In most tests biological control agents have been applied to protect susceptible crops. However, application of the fungus to a relatively poor host for the nematode on which small galls are produced so that most egg masses are exposed in the rhizosphere might provide more effective control; soil population densities of the nematode would be reduced to non-damaging levels before a susceptible crop was planted. Bridge (1987) recommends the following rotation for the management of root-knot nematodes: susceptible host-poor host-poor host-resistant or non-host-susceptible crop. Application of the fungus before the first or second poor host may permit the shortening of rotations, or the replacement of the resistant or non-host crop, without increasing nematode damage to the susceptible crops. Currently, at Rothamsted Experimental Station, United Kingdom, poor hosts for root-knot nematodes are being screened for their ability to support *V. chlamydosporium* in their rhizospheres. It is hoped that by combining the host plant and the biological control agent, more effective and consistent control can be achieved.

The Next 20 Years

The past 20 years have seen a significant increase in the number of scientists involved in research on the biological control of nematodes. Surveys and empirical tests are being replaced by quantitative experimentation and basic research on the modes of action, host specificity and epidemiology of selected organisms. Such basic information is essential for a realistic appraisal of the impact of molecular biology on the improvement of microbial agents and monitoring the spread and survival of released organisms, and for the development of rational strategies for control.

Current experience suggests that biological control agents will not replace the use of nematicides but, integrated with other control measures including chemicals, they could play an important role in the development of integrated control strategies in both developed and developing agriculture. The urgent need to reduce the dependence on nematicides should provide the necessary impetus for the considerable amount of research and development still required to ensure the successful use of such agents.

Nematode Control in the Home Vegetable Garden

Most gardeners are familiar with problems caused by diseases and insects because such problems are easily recognized. Few gardeners, however, are aware of the damage nematodes can cause.

Nematodes are microscopic (less than 1 mm long), wormlike animals too small to be seen with the unaided eye. The majority of plant parasitic nematodes live in the soil and damage plants by feeding in large numbers on the roots, impairing the plant's ability to take up water and nutrients. Severe root damage caused by nematodes typically results in aboveground symptoms that may include stunting, yellowing of leaves, loss of plant vigor and/or an overall general decline in plant performance.

Damage is often more pronounced when plants are under other stresses such as lack of water or nutrients or when damaged by other diseases or insects. Although nematodes rarely kill plants, they can drastically reduce plant growth and yields. Nematodes are usually confined to localized areas in the garden spreading very slowly under their own power; however, nematodes may be dispersed more rapidly by movement of infested soil through cultivation, on soil clinging to garden tools and tillers, in water, or on roots of transplants.

Though there are at least 10 different genera of plant parasitic nematodes found in Alabama, the most important is the root-knot nematode. Root-knot nematodes have a wide host range, but the most serious problems occur on broadleaf crops. Root-knot nematodes attack the roots of plants causing distinct knots, swellings, or galls to form on the infected roots. Galls may grow as large as one inch in diameter where they merge, but usually they are not much larger than a pea. There are several different species of root-knot nematodes in Alabama and several species may be present in any one location depending on which crops were grown previously (different species have different host ranges), sources of contamination, and geographical region in the state. Nematode species other than root-knot nematodes can cause damage to vegetables. These species include dagger, reniform, ring, stubby root, stunt, sting, root lesion, and cyst nematodes. With the exception of the cyst nematode, which produces distinctive egg-containing cysts on roots, identification of these other nematode species requires laboratory analysis. Nematode management requires long-term planning. No current control practice will permanently eradicate nematodes from the garden. Nematodes can be effectively managed in the home garden by the use of one or more of the following practices.

Site Selection

Consider nematodes when selecting a site for a vegetable garden. Have the soil from the proposed area tested for nematodes before planting. If possible, collect soil in the fall when nematode populations are highest. The worst time to sample is in late winter or early spring. Nematode populations are at their lowest during this period and may not be detected in the sample. Take samples with a soil probe or hand shovel in a zigzag pattern across the garden area from the top 8 to 10 inches of soil.

Sanitation

Sanitation aids in reducing plant parasitic nematode populations. Nematode infected plants (including roots) should be removed from the garden and destroyed as soon as the vegetables have been harvested. This practice can be particularly effective in small gardens. Plow the garden immediately after the final harvest to bring plant roots to the surface. Work the soil in this manner two to four times during the winter. The drying action of the wind and sun will destroy many nematodes and their eggs, thus preventing further buildup. Vegetable rootsleft in the soil through the winter serve as hosts on which nematodes can maintain or increase their population for the following year.

Rotating crops from year to year can be an effective means of controlling plant parasitic nematodes. Closely related crops are more likely to support the same nematodes, diseases, and other pests than unrelated plants. A particular vegetable or vegetables belonging to the same family should never be grown in the same location more than once every three years. If space is available, it is a good practice to rotate garden sites. Where garden space is limited, rotate related vegetables in one family group with vegetables in an unrelated family group. Crop rotation is not always an effective strategy for controlling root-knot nematodes because of their wide host range. However, rotating crops in combination with fallowing or soil solarization can be used to manage most nematode problems.

Use of Resistant Varieties

Resistant varieties offer the easiest, least expensive, and most effective method of controlling nematodes in the home garden. Unfortunately, resistant vegetable varieties are only available for root-knot nematodes and only for a limited number of crops. Nematodes are unable to feed on a resistant variety, resulting in a population decline over time due to starvation (as long as weed hosts of the nematodes are also eliminated). Asparagus, onion, and strawberry are resistant to most root-knot nematode populations in Alabama. For vegetable varieties resistant to root-knot nematodes. Also check seed catalogues each year to identify newly released vegetable varieties with resistance to nematodes.

Fallowing

The practice of fallowing prevents any vegetation from growing in the garden area, thus starving the nematode population. The target area must be kept completely weed free to prevent nematodes from surviving on alternate hosts. One year of fallowing will lower the nematode population to a level where an annual crop can be grown successfully. The longer an area is fallowed, the greater the decrease in the nematode population. Fallowed soil should be roto-tilled every two weeks to reduce weeds and to expose nematodes to the sun.

Plant Stress Reduction

The less stress plants are under, the better able they are to withstand nematode attack. Watering plants deeply and less frequently will encourage the development of a deep root system that will reduce stress on plants and can help minimize nematode problems. Proper management of diseases and insects can also reduce stress and help reduce damage from nematodes. Nutrient deficiencies and soil compaction can inhibit root development and increase plant sensitivity to nematode damage. Nematode damage is more severe in sandy soils than in heavy soils.

Addition of Organic Soil Matter

Organic soil amendments can improve plant health and vigor, but they must be incorporated before seeding or transplanting. Amendments that improve the soil composition, moisture retention, or physical characteristics will help produce healthier plants. Compost and pine bark are two amendments that can improve soil conditions and plant health. Pine bark, which works best if fresh and ground into small pieces, has been reported to help suppress damage from root-knot nematodes. This should be done a month or more before planting.

Suppressive Crops

Nematode suppressive crops combat nematodes naturally. Several plants minimize nematode damage in vegetable crops. These plants produce nematicidal (killing) and nemastatic (suppressive) organic compounds that are toxic to nematodes. These compounds are released from the roots of living plants or by plant's being incorporated into the soil as a green manure. Some plants may act as trap crops that prevent nematodes from maturing and reproducing once they invade the roots. Some marigolds, a few varieties of chrysanthemums, castorbeans, partridge peas, several Crotalaris spp., velvetbeans, and rapeseed are considered nematode suppressive plants. Growing a nematode suppressive crop will not eliminate nematodes from the soil; however, it may reduce nematode numbers enough to allow production of a crop in a nematode infested area. Nematode populations often rebound to pretreatment levels when a susceptible vegetable is grown following a nematode suppressive crop. Nevertheless, the use of a nematode suppressive crop has been shown to be as effective as or somewhat better than fallowing an area to reduce the nematode population.

Organic Control

Clandosan 618 produced by Igene Biotechnology in Columbia, Maryland, is an example of an organic nematicide made of chitin extracted from shellfish waste, soybean meal, and urea. The material works by increasing the soil microorganisms that feed on chitin. Chitin is one of the primary components in the body wall of nematodes. The feeding activities of the chitin-loving microorganisms have been shown to reduce the number of nematodes for two cropping sequences. Control is sporadic, however, and requires large quantities of an expensive product. Chitin products should be incorporated into the soil about two weeks before planting.

Chemical Control

There are no nematicides on the market that can be used by homeowners. Currently, licensed pesticide applicators can still use metam sodium, a preplant fumigant, on home gardens, but this chemical requires important safety precautions and may carry significant legal liabilities. Therefore, chemical control is not a viable option in home vegetable gardens.

Soil Solarization

Soil solarization is a simple, safe, and effective method of nematode control. It allows the grower to bypass lengthy crop rotations and gives the added benefit of controlling other soil pests such as insects and weeds. Radiant heat from the sun is the lethal agent involved in soil solarization. A clear polyethylene mulch or tarp is used to trap solar heat in the soil. Over a period of several weeks to a few months, soil temperatures become high enough to kill nematodes, as well as many other soil pests and weed seed to a depth of 8 inches. None of the pests will be completely eradicated, but their numbers will be greatly reduced, allowing successful production of a crop. In sandy or sandy-loam soils, nematodes may survive at depths below the lethal temperature zone. As a result, some damage may be seen on deep-rooted crops, but those with shallow root systems should escape serious injury.

The soil to be solarized must be cultivated until it is loose and friable with no large soil clods or debris. A roto-tiller will eliminate clods or other debris creating air pockets that reduce heating of the soil or keep the tarp from fitting tightly over the soil surface. Make sure moisture levels are adequate for working the soil before laying the plastic tarp. Use a clear, UV-stabilized plastic tarp or sheeting 1 to 4 millimeters thick. The edges of the tarp must be buried to a depth of 6 inches in the soil to prevent blowing or tearing by the wind. White or black plastic usually does not transmit enough solar radiation to raise soil temperatures to lethal levels for many soil pests. Long, hot sunny days are needed to reach the soil temperatures required to kill soil pests and weed seed. The longer the soil is heated, the better and deeper the control of nematodes and other soil pests will be. In Alabama, a tarping period of 4 to 6 weeks in the summer should be sufficient to kill nematodes and soilborne plant pathogens.

Introduction to Plant-Parasitic Nematodes

The History of Nematodes

Members of the phylum Nematoda (round worms) have been in existence for an estimated one billion years, making them one of the most ancient and diverse types of animals on earth. They are thought to have evolved from simple animals some 400 million years before the "Cambrian explosion" of invertebrates able to be fossilized. The two nematode classes, the Chromadorea and Enoplea, have diverged so long ago, over 550 million years, that it is difficult to accurately know the age of the two lineages of the phylum.

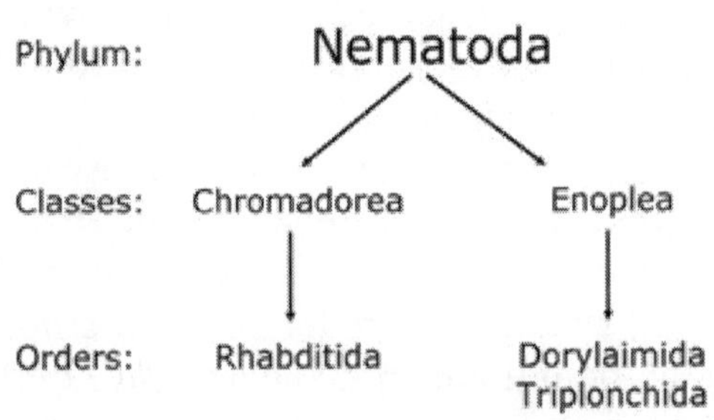

Nematodes are multicellular animals in the group Ecdysozoa, or animals that can shed their cuticle. Also included in this group with nematodes are insects, arachnids and crustaceans. In contrast to some of their relative invertebrates, nematodes are soft-bodied. Thus, very few nematodes have been fossilized (22 species from 11 genera) and exactly what ancestral nematodes looked like remains unknown. While we do not know the morphology of the first nematodes, it is probable that they were microbial feeders in the primordial oceans. The oldest known fossil nematodes are only 120-135 million years old; by then nematodes had diversified to feed on microbes, animals and plants. The oldest fossil nematodes are found in amber and are commonly associated with insects. This is probably due to the fact that tree sap, which fossilizes to make amber, captures and preserves insects and their associated nematodes much more easily than an animal-or a nematode-infested portion of a plant. Much of what we know about the evolution of nematodes is inferred from the comparative anatomy of existing nematodes, trophic habits, and by the comparison of nematode DNA sequences. Based upon molecular phylogenic analyses, it appears that nematodes have evolved their ability to parasitize animals and plants several times during their evolution. One point is clear; nematodes have evolved to fill almost every conceivable niche on earth that contains some amount of water. Nematodes are extremely abundant and diverse animals; only insects exceed their diversity. Most nematodes are free-living and feed on bacteria, fungi, protozoans and other nematode (40% of the described species); many are parasites of animals (invertebrates and vertebrates (44% of the described species) and plants (15% of the described species).

Nematodes were noted early in human history because some serious human diseases are caused by relatively large vertebrate-parasitic nematodes. Some of these nematodes were first described in the ancient Chinese scientific literature as early as 2700 B.C.. Since plant parasitic nematodes often are small and subterranean, there are not many ancient references to phytoparasitic nematodes. One interesting observation suggests that phytoparasitic nematodes were known in antiquity (235 B.C.) because the ancient Chinese symbol for a soybean root-infesting organism resembles in shape an adult female soybean cyst nematode. The first described plant parasitic nematodes were discovered in wheat seeds by Needham (1743). Not until the identification of root-knot nematodes on cucumber by Berkeley (1855) and cyst nematodes causing "beet-tired" disease on sugar beets by Schacht (1859), did plant nematology begin to emerge as an important scientific discipline. Nathan A. Cobb, the "father of US nematology," pioneered agricultural nematology as a USDA scientist in the early 1900's. The use of soil fumigation to reduce nematode populations and increase crop yields in the 1940's (Carter) demonstrated that nematodes were significant crop pathogens and ushered in the "chemical era" for nematode management in production agriculture. Today plant parasitic nematodes are recognized as major agricultural pathogens and are known to attack plants and cause crop losses throughout the world. Some estimates suggest they cause 77 billion dollars of damage worldwide each year. As the full extent of damage caused by plant-parasitic nematodes is recognized by agricultural scientists, the study of the biology of plant-parasitic nematodes will become increasingly important.

Nematode Biology

The study of nematode biology has led to a dramatic increase in understanding of how all animals function. In fact, the bacterial-feeding nematode, *Caenorhabditis elegans*, is one of the best-understood animals on earth. The fate of every cell in *C. elegans* development has been carefully mapped. It was the first animal to have its DNA sequence completely deciphered, and it is amenable to detailed genetic analysis. The study of *C. elegans* has led to new insights into the details of animal development, neurobiology, behaviour, and has been of great value in biomedical research as well as in the understanding of nematode biology.

The reason nematodes are so useful for biological research is due to their simple anatomy and transparent bodies. Nematodes are simple animals, often only containing 1000 cells or less. Nematodes in all or part of their life cycle are worm-shaped (vermiform), although some species become swollen and rounded in later life stages. The basic body plan of a nematode is a tube within a tube. They have an outer skin or cuticle that is secreted from an inner hypodermis.

The muscles are attached longitudinally to the nematode's hypodermis, allowing them to move only in the dorsal ventral direction (snake-like movement). Inside the nematode there is an inner tube, the alimentary canal, which runs inside the nematode from head to tail. Between the alimentary canal and the body wall is fluid that provides pressure against the wall to maintain body shape and allow movement. At the head of a plant-parasitic nematode is a hollow mouth spear (like a hypodermic needle) called a stylet. The nematode uses this stylet to puncture plant cells, to withdraw food and also to secrete protein and metabolites that aid the nematode in parasitizing the plant. The stylet is connected to the pharynx that, in turn, is connected to the intestine. The intestine ends at the rectum in the female nematode and the cloaca in the male. Attached to the pharynx are three-five salivary glands which produce secretions that may be emitted from the stylet and that assist the nematode in plant invasion and parasitism.

The nematode pharynx is muscular and specialized areas can contract and expand the esophageal lining. The expansion and contraction of the pharynx muscles allow the nematode to pump food into its intestine through its stylet or eject secretions from its salivary glands into and around plant cells. In rhabditidan nematodes, the esophageal muscles are localized in a muscular organ called the metacorpus. In enoplean nematodes, the esophageal muscles are more spread out and do not form a compact pumping organ.

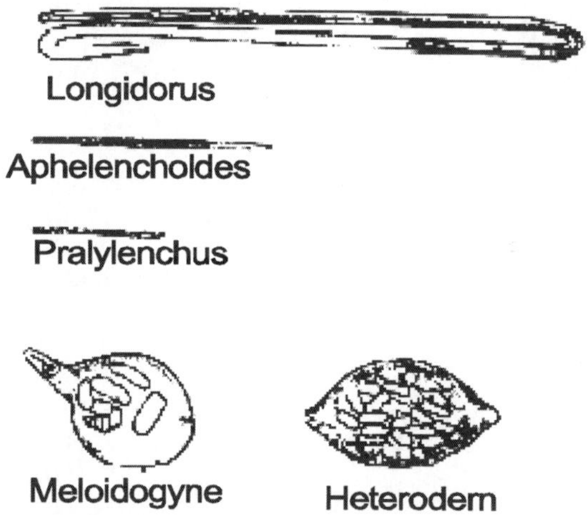

In the middle to posterior of the nematode are the reproductive organs. Nematode species often have both males and females, but it is not uncommon for plant nematodes to reproduce asexually by parthenogenesis. In females the reproductive organs are used as traits for identification because the number of ovaries and the position of the vulva in the female nematode's body are easily seen under the light microscope. Male nematodes have one or two testes and they are easily identified by the presence of spicules. Spicules are copulatory structures that are used during mating to guide the sperm into the vagina of the female nematode.

The common plant-parasitic nematode genera are fairly easy to identify to that level using a standard compound microscope. Identification of nematodes to the species level often requires detailed morphological analysis, growth of the nematode on different host plants, or DNA or isozyme analysis. Common morphological features used in nematode identification include the mouth cavity (presence or absence and shape of a stylet), the shape and overlap of the pharyngeal glands with the intestine, size and shape of the nematode body at the adult stage, size of the head, tail, and number and position of ovaries in the female. More subtle characters may include number of lines on the nematode's cuticle or the presence or absence of pore-like sensory organs. An excellent guide to nematode identification is "Plant-Parasitic Nematodes: a Pictorial Key to Genera" by William F. Mai et al., 1996.

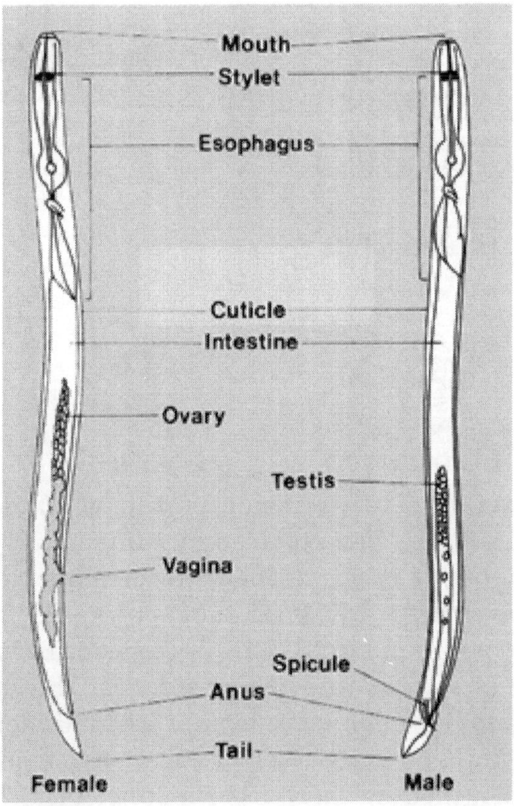

Plant-parasitic nematodes occur in all sizes and shapes. The typical nematode shape is a long and slender wormlike animal, but often the adult animals are swollen and no longer even resemble worms. Plant-parasitic nematodes range from

250 um to 12 mm in length, averaging 1 mm, to about 15-35 um in width. While nematodes may look dramatically different, they all share some common features. Nematodes often look segmented because of the numerous annulations (accordion-like transverse grooves) on the cuticle that allow the nematode to bend without kinking, but in fact nematodes are unsegmented and have no replication of body parts throughout the worm.

Like most higher animals nematodes possess bilateral symmetry, but with a superimposed trilateral and hexalateral symmetry. Developmentally, nematodes are triploblastic, containing three body layers (ectoderm, mesoderm and endoderm) in the embryo. Most higher organisms are tripoblastic and have a coelom, a body cavity surrounded by mesoderm. Nematodes have a body cavity that is not totally surrounded by mesoderm, so they are pseudocoelomic.

Nematodes are evolutionarily related to insects, and one feature they have in common is the requirement to molt between juvenile stages. All nematodes undergo four molts from the juvenile to the adult phase of their life cycle. They have four juvenile stages and an adult stage. In many nematodes the first molt usually occurs in the egg and it is the second-stage juvenile that hatches. While all nematodes undergo four molts, molting is not required for growth of the nematode as it is in insects because of the elasticity of the nematode cuticle. Nematodes do not have a skeleton, but they do have a hypodermis which functions as a flexible support for their muscles.

Nematode muscles are arranged along the longitudinal axis of the worm, thus allowing the body of the nematode to move only in a sinusoidal (snake-like) motion. Nematodes have no defined respiratory or circulatory systems; they depend on diffusion of water, gasses and metabolites in and out of their semi-permeable body walls and internal transport by mixing of the pseudocoelomic fluid as the nematode moves. This lack of respiratory/circulatory systems prevents nematodes from becoming larger in cross sectional area, but does not limit their length. The largest nematode found thus far was more than 7 meter long and 1 cm in diameter. Sorry, no giant nematodes; they are only in the movies. Nematodes have a sophisticated nervous system and sensory organs to help them find their host plant, to locate specific plant cell types, and to mate and reproduce.

Nematode Survival Strategies

One might think of soil as a safe environment, but to a microscopic nematode it is a hostile world filled with danger. A nematode must contend with voracious predators, changes in soil temperature and moisture, and the death of its host plant. For a nematode population to survive, it must be able to circumvent these obstacles. Nematodes evade these biotic and abiotic obstacles by employing a combination of behavioural and physiological survival strategies.

While all nematodes feed on other organisms, the soil is filled with bacteria, fungi and other nematodes that would gladly consume a nutrient-rich plant-parasitic nematode. In fact, the study of nematode predators and pathogens is an important area of study in nematology because nematode predators can be used (in theory) to control populations of plant-parasitic nematodes. While nematodes do possess a thick cuticle that may provide some protection from predation, this type of defence is easily breached by specialized nematode pathogens.

The most common method plant nematodes use to evade predation is by living inside plant tissue or by limiting their mobility in the soil environment. By spending less time moving in the soil, a nematode can reduce its chance of "running into" a predator or pathogen. Some plant nematodes spend most of their time in the soil (ectoparasites) and others are mostly contained within the plant tissue (endoparasites). Nematodes that live inside plants have some degree of protection from predation, but they risk death if their host plant succumbs to disease. In contrast, nematodes that move from host to host reduce the risk of perishing with their host, but have a greater chance of encountering a predator or pathogen.

Nematode survival is not impacted only by biotic factors, but also by abiotic ones such as temperature and water availability. The onset of winter or the drying of the soil can be disastrous for a nematode. Interestingly, many nematodes are well adapted to abiotic stress and are capable of cryptobiosis (hidden life): the ability to enter a state of suspended metabolic activity during unfavourable environmental conditions (drying, heat, cold). While not all nematode are capable of cryptobiosis, the ones that are can often survive for years in a cryptobiotic state awaiting favourable conditions that will trigger their revival. The ability of nematodes to undergo cryptobiosis is one reason some nematode species are very difficult to eradicate from a field.

Although cryptobiosis is a useful survival strategy, not all nematodes can enter such a survival stage. Instead, some nematodes have adapted to feed upon many different plant species, thus avoiding the disaster of losing a host plant. Nematodes with a broad host range can feed upon weeds and wild plants making their eradication virtually impossible. By using a complex network of behavioural and biochemical protections nematodes are able to survive and prosper in what would seem to us to be hostile environments.

Nematode Dissemination

While nematodes are motile animals, most are able to move no more than a meter through the soil within their lifetime. However, this lack of long distance crawling does not mean nematodes cannot rapidly spread from field to field. Farm equipment and even muddy shoes contaminated with nematode-infested soil can rapidly disperse nematodes.

The movement of water during floods and irrigation can disperse nematodes over long distances. Likewise the movement of nematode infected plants, seeds, and bulbs can give nematodes international tickets to travel the world if plant quarantine officials are not careful. The ability of nematodes to form environmentally resistant stages makes their dissemination even easier, since dried nematodes can be blown with the wind or plant debris over large geographical regions. Even migrating birds are suspected to be able to carry nematodes along their flight paths, assisting the nematodes in their quest for new homes. Essentially any process that moves soil or plant tissue has the ability to disperse plant nematodes, making them difficult plant pathogens to quarantine.

Plant-Nematode Interactions

Nematodes feed on all parts of the plant, including roots, stems, leaves, flowers and seeds. Nematodes feed from plants in a variety of ways, but all use a specialized spear called a stylet. Note the differences in stylet length and shape. The size and shape of the stylet is used to classify nematodes and also can be used to infer their mode of feeding. All three nematodes in figure below are ectoparasites, but *Belonolaimus* and *Longidorus* feed deep within the roots using their long stylets, while *Helicotylenchus* feeds on the exterior of the root or partially burrows into the root to feed using its short stout stylet.

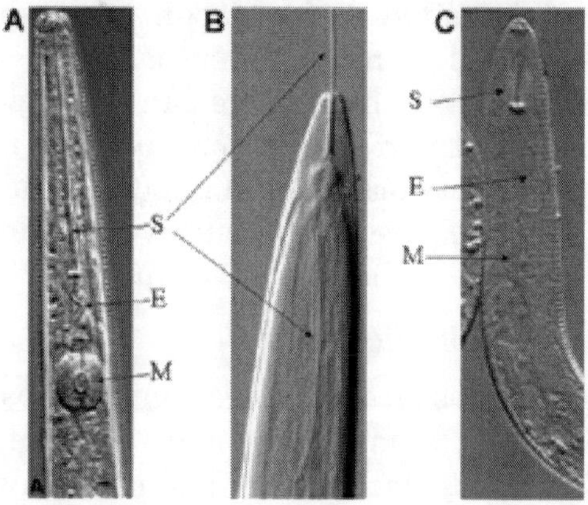

Often nematodes withdraw the contents of plant cells, killing them. When this type of feeding occurs, large lesions are formed in the plant tissue. Some nematodes do not kill the plant cells they feed upon but "trick" the plant cells to enlarge and grow, thus producing one or more nutrient-rich feeding cells for the nematode. These feeding cells enable long term feeding associations, and form by repeated nuclear division in the absence of cell division (giant cells) or by the incorporation of adjacent cells into a syncytium formed by the breakdown of neighbouring cell walls. Collectively, nematodes can feed on almost any plant cell type, and form a variety of feeding cell types. The number of feeding cells can vary from one to a half dozen depending on the nematode species.

Many plant-parasitic nematodes feed on the roots of plants. The feeding process damages the plant's root system and reduces the plant's ability to absorb water and nutrients. Typical nematode damage symptoms are a reduction of root mass, a distortion of root structure and/or enlargement of the roots. Nematode damage of the plant's root system also provides an opportunity for other plant pathogens to invade the root and thus further weakens the plant. Direct damage to plant tissues by shoot-feeding nematodes includes reduced vigor, distortion of plant parts, and death of infected tissues depending upon the nematode species.

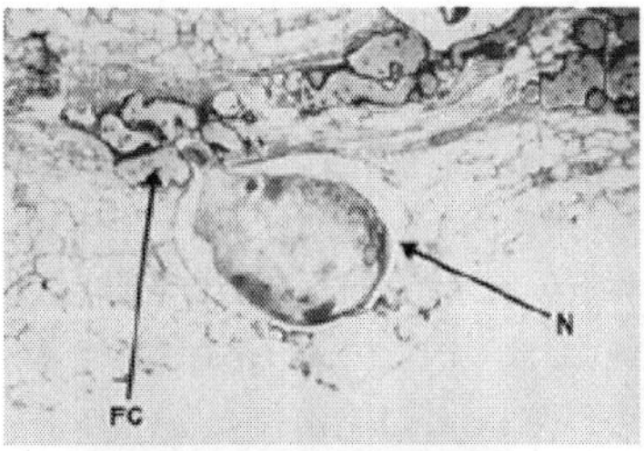

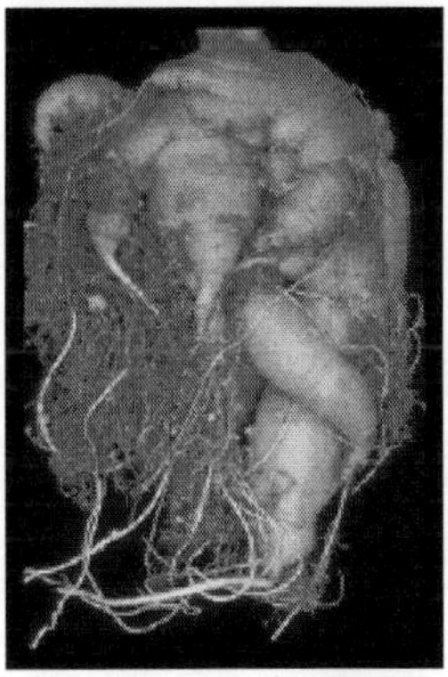

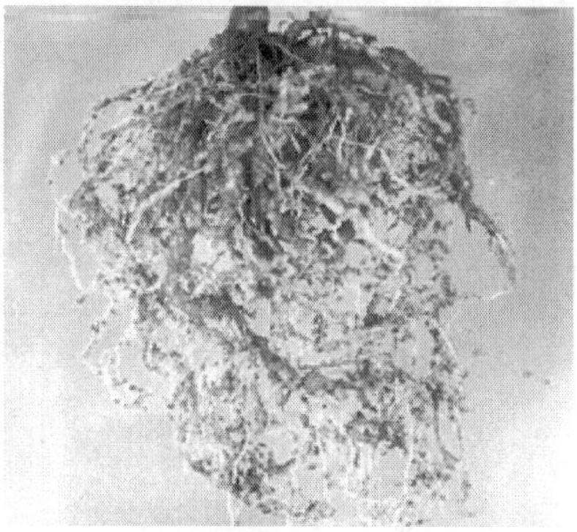

The above ground symptoms of nematode damage to roots are relatively nondescript, including nutrient deficiency, incipient wilt, stunting, poor yield and sometimes plant death. Few diagnostic signs and symptoms of plant damage by nematodes exist except root galls, cysts, "nematode wool," and seed galls. Thus, damage to crops by root-infesting nematodes often goes unnoticed by growers. Field patterns of nematode damage to roots begin in a small area and spreads radially from the initial infection site, often assisted by farm equipment. The only way to accurately diagnose nematode disease is to sample soil and plant material from suspected sites and extract nematodes for analysis. Nematodes are extracted from the soil by floating them in water to remove heavy soil particles and then catching the floating nematodes on sieves with fine pore sizes. Plant tissues infected with motile nematodes can be incubated in a Baermann funnel or moist chambers to collect nematodes that will exit the tissues.

Most plant parasitic nematodes are soilborne root pathogens, but a few species feed primarily upon shoot tissues. The majority of plant parasitic nematode species are in the class Chromodorea, order Rhabditida (formerly placed in the order Tylenchida). There are seven major types of nematode feeding strategies used by plant parasitic nematodes.

Ectoparasites: The first feeding type is the ectoparasitic mode, in which the nematode remains outside of the plant and uses its stylet to feed from the cells of the plant roots. Nematodes that use this strategy can graze on numerous plants, making it easier for them to switch hosts, but their added mobility makes them very susceptible to environmental fluctuations and predators. Ectoparasitic nematodes can have extremely long stylets, which assist them in feeding deep within the plant root on nutrient rich plant cells. Some of these nematodes induce the plant to form an enlarged cell(s) that the nematode feeds from for an extended period of time. Note, in all life cycle diagrams in this article the abbreviation J=juvenile and the number refers to the stage of the nematode and M=molt and refers to how many molts the nematode has completed.

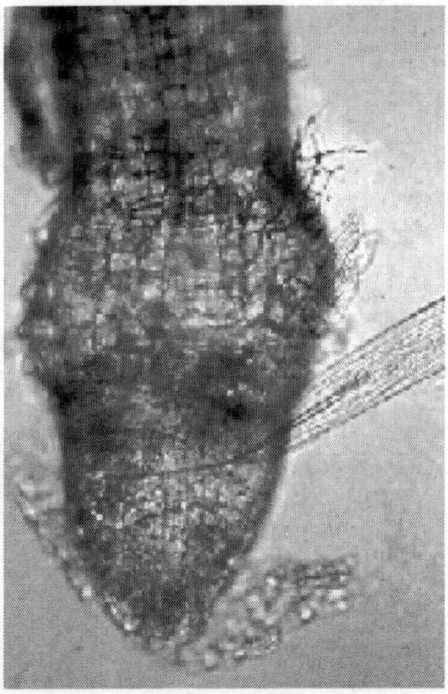

The life cycle is typical for a nematode in the class Enoplea, but most nematodes in the class Chromadorea undergo their first molt in the egg and hatch as J2. All motile nematode stages are capable of feeding from the plant. The nematodes feed, undergo four molts into adults, mate and lay eggs. Some nematodes that use this feeding strategy can form terminal galls in the roots and cause severe stunting of the root system. An example of an ectoparasite is *Xiphinema* (dagger nematode), shown below, feeding from a fig root. This enoplean nematode is particularly problematic because it can harbour and transmit plant viruses. Only ectoparasitic nematodes in the class Enoplea transmit viruses, but because of the virus transmission, this type of nematode in low numbers can still be very damaging to plants.

Semi-endoparasites: Nematodes that feed as semi-endoparasites are able to partially penetrate the plant and feed at some point in their life cycle. Usually the head of the nematode penetrates into the root and allows the nematode to form a permanent feeding cell. These nematodes swell and do not move once they have entered into the endoparasitic phase of their life cycle. By giving up their mobility, the nematodes risk death if their host plant dies, but they also benefit from forming a permanent feed site, which increases their nutrient uptake and reproductive potential.

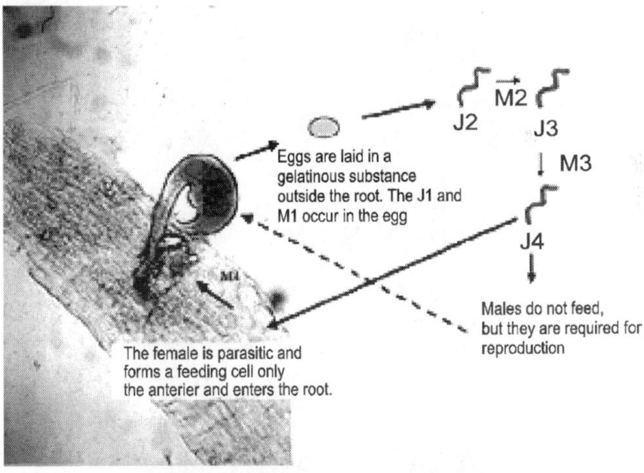

A typical nematode with this life cycle is *Rotylenchulus reniformis*, the reniform (kidney-shaped) nematode. This nematode hatches from the egg as a J2, then quickly molts in the soil to the adult stage without feeding. The anterior end of an adult female enters the plant root and forms a feeding cell. After mating, the female lays its eggs outside of the root in a gelatinous egg mass Maggenti (1981). Another nematode with a similar feeding strategy is *Tylenchulus semipenetrans*, the citrus nematode, although the juvenile stages of this nematode do feed as ectoparasites. As is common in biological systems, it is often difficult to precisely classify animals due to variation in their behaviour. True to this rule, several species of ectoparasitic nematodes (e.g. *Helicotylenchus*) are also capable of partially penetrating the root and feeding. However, we do not classify these nematodes as semi-endoparasites because they do not exhibit a consistent endoparasitic feeding behaviour.

Migratory Endoparasites: Migratory endoparasitic nematodes spend much of their time migrating through root tissues destructively feeding on plant cells. These nematodes cause massive plant tissue necrosis because of their migration and feeding. When they feed from the plant, they simply suck out the plant cell cytoplasm using their stylet, killing the plant cell and moving ahead of the lesion. They make no permanent feeding cells. In a typical life cycle, the nematode hatches from the egg as a second-stage juvenile and starts feeding on the plant.

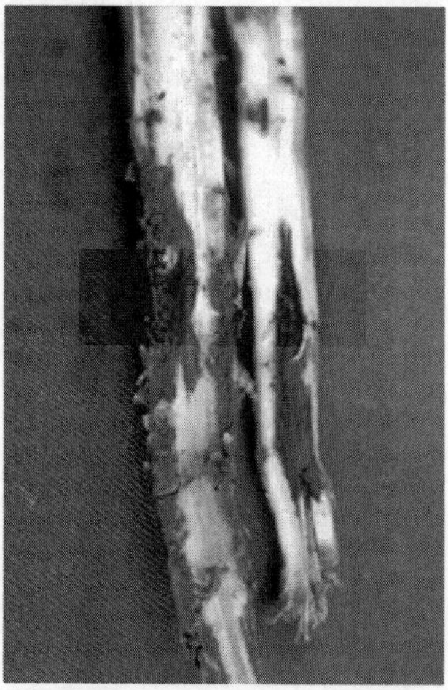

The nematodes feed, molt and reproduce primarily within the plant tissue. All motile stages are capable of feeding from the plant and they are able to move into the soil in search of new roots to invade. Because these nematodes create extensive wounds in the plant root, secondary infection by bacteria and fungi can often occur, further damaging the root system. Below is a typical life cycle for a migratory endoparasite. Examples of migratory endoparasitic nematodes are *Pratylenchus* (lesion nematode), *Radopholus* (burrowing nematodes) and *Hirschmanniella* (rice root nematode).

Sedentary Endoparasites: The most damaging nematodes in the world have a sedentary endoparasitic life style. The two main nematodes in this group are the cyst nematodes (*Heterodera* and *Globodera*) and the root-knot nematodes (*Meloidogyne*). In these nematodes, the J2 invades the plant near the tip of a root and migrates through the tissue to the developing vascular cells. These nematodes are completely embedded in the root during their initial stages of development, but later the cyst nematodes protrude from the root. The J2 nematodes inject secretions into and around the plant cells to stimulate the formation of large feeder cell(s), which they non-destructively feed on throughout their life cycle. Feeding cells of root-knot nematodes (giant cells) form by repeated nuclear division in the absence of cell division. Feeding cells of cyst nematodes form by the incorporation of neighbouring cells into a syncytium formed by the breakdown of neighbouring cell walls. Once the feeding cells are formed, the juvenile nematode rapidly becomes sedentary because their somatic muscles atrophy. The juveniles feed, enlarge and molt three times to the adult stage. The large feeding cells formed by these nematodes plug the vascular tissue of the plant making it susceptible to water stress. Female sedentary endoparasites enlarge considerably into a saclike shape and are capable of laying large numbers of eggs. Eggs are typically laid outside the nematode in a gelatinous egg mass, but in cyst nematodes most eggs are retained inside the female body. Both types of nematode have the same basic feeding strategy, but many cyst nematodes have an obligate sexual cycle, whereas the most common species of root-knot nematodes reproduce largely by parthenogenesis.

Below, the life cycle of the soybean cyst nematode (*Heterodera glycines*) is described. The cyst nematodes are extremely problematic because they have the ability to persist for a long period of time in a field. Their remarkable persistence is due to their ability to produce a cyst, which is the hardened dead body of the female nematode that surrounds the eggs.

The eggs in the cyst are in a dormant state and they hatch slowly over years in the field, usually in response to signals from host roots. Having a resistant or dormant nematode stage enables nematodes to survive non-optimal conditions. Resistant stages also aid the nematode in dispersal. While the root-knot nematodes do not have an environmentally resistant stage, they do possess a broad host range that enables them to persist on an alternate host. The use of environmentally resistant nematode stages and broad host ranges are two key strategies plant nematodes use to persist in the environment.

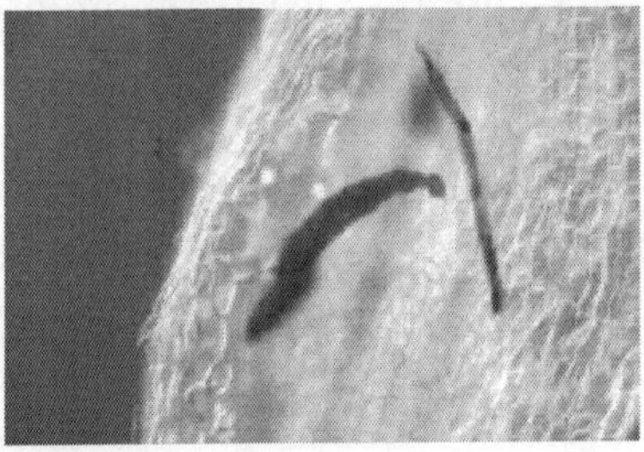

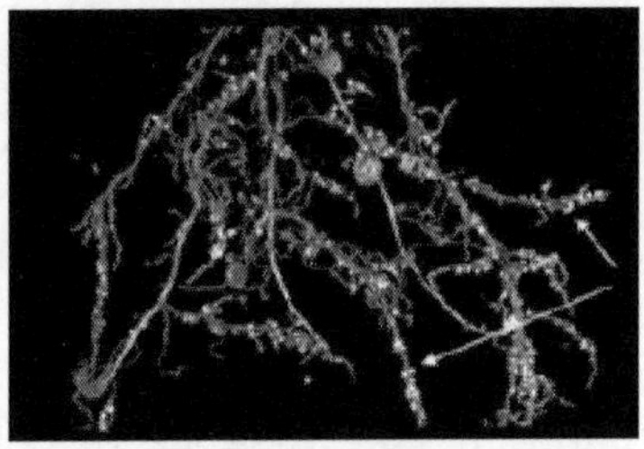

Stem and bulb nematodes: Stem and bulb nematodes (*Ditylenchus* spp.) are, as their name suggests, nematodes that attack the upper and lower parts of plants. They use water films to migrate up the stem of the plant and therefore are more damaging under wet conditions. The infectious stage of the stem and bulb nematodes is the fourth stage juvenile.

Stem and Bulb Nematodes

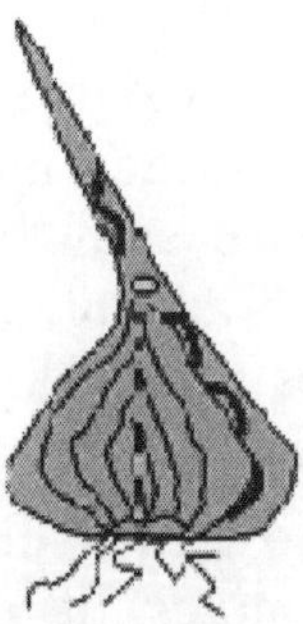

This stage often enters emerging plant tissues below ground, but can crawl up stems in a film of water and enter shoots via buds, petioles, or stomata. Once in the host plant, they destructively feed as migratory endoparasites, molt into adults and reproduce. The nematodes hatch from the egg as J2 and continue to feed, molt and reproduce, extensively macerating and distorting the plant tissue. Once the plant is destroyed or winter arrives, the stem and bulb nematode juveniles arrest their development at the environmentally resistant J4 stage and overwinter. Fluffy masses of dried (cryptobiotic) *Ditylenchus* can be seen on the surface of bulbs and are known as "nematode wool." Once environmental conditions are favourable, the cryptobiotic J4 become active and their life cycle resumes.

Bursaphelenchus xylophilus (pine wood nematode), a nematode that infects pine trees, has a different and fascinating life cycle. Resistant stages of these nematodes are carried into pine tree in the tracheae of bark beetles. Once in the tree the nematodes are released from the beetle, and they migrate through the resin canals of the tree.

Pine wood Nematode

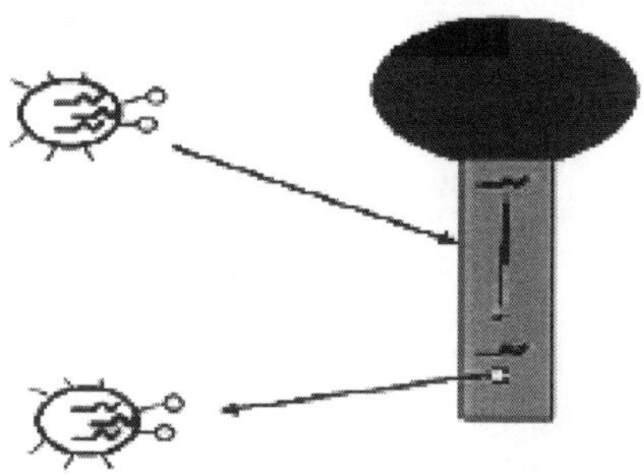

The pine wood nematode destructively feeds from cells in the resin canals, clogging them and the tree's vascular tissue causing pine wilt disease. The nematodes feed, molt and sexually reproduce at a highly accelerated rate. The population of pine wood nematodes builds up rapidly and quickly kills the host tree. The nematodes are able to feed on fungi in the tree after the plant cells are dead. The nematodes survive the winter as J3s and are attracted to beetle larvae. In the spring the nematodes molt to the resistant J4 stage and invade the beetle to be dispersed to neighbouring trees. The ability of the nematode to be dispersed by an insect allows them to spread rapidly from tree to tree, a process which can quickly destroy a forest. This nematode is thought to be indigenous to the Unitied States where all of the native trees are resistant to the damaging effects; however in Asia and Europe, the pine tree species are very susceptible, which has resulted in the death of millions of trees each year.

Seed gall nematodes: Seed gall nematodes (*Anguina* spp.) were the first plant-parasitic nematodes to be described in the scientific literature in 1743. These nematodes migrate as J2s in water films to the leaves of plants where they feed as ectoparasites at the tips, causing distortion of the leaves. Once the plant starts to flower the J2 penetrates the floral primordia and starts to feed on the developing seed. Once in the seed, the nematode undergoes its molts, continues to feed, and eventually kills the seed to form a blackened "cockle" (seed gall). The adults sexually reproduce, the eggs hatch as J1 and then quickly molt into a J2 survival stage. The environmentally resistant J2 desiccates with the seed gall and overwinters. The nematodes in the seed gall can survive for 30 years if kept in a dry location. When proper moisture and temperature conditions arise, the cryptobiotic J2 becomes active and starts the life cycle over again.

Seed Gall Nematodes

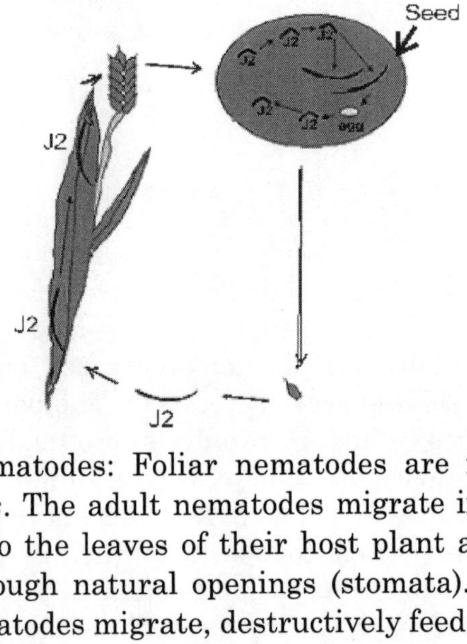

Foliar nematodes: Foliar nematodes are in the genus *Aphelenchoides*. The adult nematodes migrate in water films on the stems to the leaves of their host plant and penetrate the leaves through natural openings (stomata). Once in the leaves the nematodes migrate, destructively feed, molt and lay eggs. The feeding activity of the nematodes causes characteristic interveinal chlorosis and necrosis of the leaf, ultimately killing it. The nematodes are able to move from leaf to leaf if the proper (moist) environmental conditions exist and can severely damage a plant. In the winter the adult nematodes persist in the dead leaves until favourable conditions arise in the spring. If the dead, nematode-infested leaves are moved or blown around this will help disperse the nematode near new host plants.

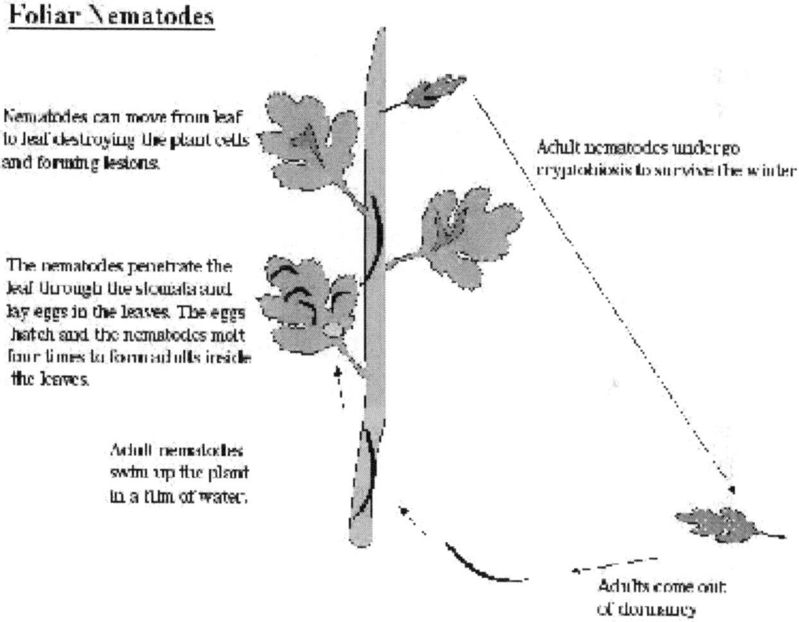

Plant Nematode Control

There are several methods commonly used to control plant-parasitic nematodes. These methods can be divided in to three main types: biological control, cultural control and chemical control. The most practical form of biological control is the use of nematode-resistant plants. In this control method, plant breeders cross natural nematode resistance genes into cultivated plant species to improve their resistance to nematodes. The benefit of this method is that it is a very inexpensive way for growers to control their nematode problems. The main disadvantage is that it takes years to screen for resistant plant varieties and more time to breed resistance traits into commercial varieties. Further complications are that natural sources of nematode resistance do not exist for all cultivated species and some species of nematodes are able to grow on resistant plants.

However, when "good" resistant plants are available, they are an effective method of nematode control. To find a list of plants resistant to nematodes see the plant host database web site in the Department of Nematology at the University of California at Davis. Biotechnology has the potential to produce nematode resistance genes that would be effective against many types of nematodes and would function in many different plant species.

However, at the present time there are no plants with bioengineered nematode resistance available for commercial use. Other biological control methods use natural predators or pathogens of nematodes. While biological control agents kill nematodes in controlled laboratory settings, implementation of biological control methods in the field is not done due to the expense and difficulty of growing large amounts of nematode pathogens. However, if some of the difficulties associated with growing nematode pathogens can be overcome, this may be a promising method of nematode control in the future.

Crop rotation with a non-host plant is a very effective method to limit nematode growth. Typically, a cropping system is devised that selects plants that nematodes can and cannot grow on. These plants are grown in alternate years and the problematic nematode population decreases dramatically, below damage threshold levels, in the years that the non-host is grown. This can be an effective method if a producer has the choice of several different crops that can be grown and if the problematic nematode does not have a broad host range or survive in the soil in a cryptobiotic state for long periods of time.

For the past 50 years nematodes have been effectively controlled using chemical nematicides. These are inexpensive chemicals that effectively kill nematodes in soil. There are two types of nematicides, soil fumigants (gas) and non-fumigants (liquid or solid). Soil fumigants became popular because they did not rely on alternative host crops for rotation; they drastically reduced nematode populations in the soil, and were cost effective for most crops. Most fumigant nematicides have been banned by the EPA as environmental toxins with the exception of 1,3 dichloropropene, chloropicrin (tear gas), and dazomet (Basamid). The multipurpose soil fumigant methyl bromide also provides excellent reduction of soil nematode populations, but methyl bromide was largely discontinued in 2005.

Non-fumigant nematicides such as fenamiphos (Nemacur) and aldicarb (Temik) are based upon the same kinds of active ingredients as many insecticides (i.e. nerve poisons) and can be applied in liquid or granular formulations. While non-fumigant nematicides reduce nematode populations, their effectiveness is not as consistent as that of fumigant nematicides. The EPA is also restricting the use of non-fumigant nematicides. Since nematicides are expensive to develop, new ones are rarely released on the market today. While nematicides are effective in controlling nematodes, they are only practical for use on high-value crops.

Chapter 10

Nematode Species Delimitation

The goal of this workshop, held in Tucson, Arizona in July 1997 and funded by the National Science Foundation (DEB 96 26813), was to bring systematists and ecologists representing terrestrial, invertebrate, vertebrate, marine and freshwater habitats of nematodes, together with systematists studying other poorly known organismal groups. The participants consisted of undergraduates, graduate students, technicians, and agency and academic scientists from the USA, Canada, Belgium and Australia.

Globally, habitats are being changed for human use, with loss of species and change in species communities. Nematodes are among the most numerous and speciose animal taxa, are a major constituent of earth's biodiversity, occurring in a majority of global habitats, and are of critical importance in the functioning of ecosystems. Issues of biodiversity science related to discovery of new species, to systematics, ecology, medicine, conservation biology, pest management, invasive species and biodiversity prospecting are only a few reasons for conducting nematode surveys and inventories. Because nematodes are microscopic roundworms that inhabit a variety of opaque substrates, the taxonomically diverse species comprising a nematode assemblage in soils, freshwater and marine sediments are typically surveyed or inventoried together, whereas nematode parasites of invertebrates, vertebrates and plants are typically targeted in surveys and inventories. It has been increasingly recognized by ecologists and other scientists that nematodes are among the most poorly known groups of all biota.

A major reason for the neglect of nematodes concerns difficulties with their adequate characterization and identification. Their simple anatomical design lends itself to convergence and may make description of new species difficult. Few animal phyla have been so resistant to mastery. As a result, few organismal groups stand to benefit more significantly than nematodes from the combination of morphological, molecular, and biochemical approaches to surveys and inventory.

Recovering all of the evolutionary units that reflect the epistemologically real species in a given habitat can be confounded by the constraints inherent to conducting a survey or inventory. Principal investigators proposing to survey or inventory nematode species must addresss the issue of which species concept they will adopt, and make explicit the operations they will use to most accurately delimit these species. These operations must be suited to the geographic/ecological scope of the study and to its time constraints. Further, they should not preclude the kinds of in-depth follow-up studies that a survey or inventory will inevitably generate. A necessary first pass will assign nematodes to morphotypes, since in many and perhaps most cases the nematodes will be assignable to some taxonomic level (family, genus) but not to named species. Morphotypes should then be subjected to further analysis if possible. If the survey's objective is not to analyse to species level, the animals should be preserved in a suitable manner and sent to authorities for archiving for future analyses.

The exact nature of further analysis may vary among studies, but principal investigators must justify their choice of analyses within the context of their species concept. Thus, the principal investigators will articulate the level of difference they accept as defining a species with reference to:

- The species concept they advocate as most accurately recovering nematode species.
- The operational ability of the biochemical, molecular or other analysis to delimit species as defined by the species concept advocated.
- The goals and scale of the proposed study.
- The potential use of data by the principal investigators and other researchers.

Discovery of New Nematodes

The rationale for conducting a nematode survey or inventory should place due emphasis on the discovery of new nematode species. The newly discovered nematodes may be important in different habitats for different reasons, and as a result, a survey or inventory involves the consideration of taxonomy, habitat, and scientific and economic importance. These issues are summarized below. Because methods of nematode extraction recover entire assemblages from soil, a survey or inventory will have to be focused to some degree, perhaps by emphasizing a taxonomic group, microhabitat, or ecosystem service. However, the focus of a survey or inventory should not result in discarding non-target nematodes. Provision should be made for the passing on of the nematodes that are outside the project's immediate interests and expertise. This is essential to obtain maximum value of the survey or inventory.

Taxonomic considerations. We want to recognize nematode groups that are species-rich but under-represented in existing taxonomic collections, i.e., Enoplia, Chromadorida and Dorylaimida. An additional consideration is the discovery of significantly novel taxa representing new families or genera.

However, any push to discover new nematodes must be tempered by extant taxonomic knowledge. The discovery process must be grounded in the reliable diagnosis of known species. Hence, new data may well need to be gathered on poorly described known species. In particular, molecular and/or biochemical characterization must not be applied only to specimens from the surveyed target area, but also (to the degree possible) to relevant animals from type localities and/or reference cultures originally obtained from outside the target area.

Habitat considerations. Most non-agricultural habitats, including marine and freshwater habitats have been poorly sampled for nematode taxa and so are of high priority for the discovery of new species. Of related interest are ecologically important habitats, such as remnant native forests and habitats that are under threat of extensive human impact, such as freshwater streams, lakes and wetlands. Accordingly, geographic location should be considered when devising a sampling protocol: as a rule-of-thumb, widely-spaced samples from within a habitat-type have priority over those in close proximity.

Scientific and Economic Importance: The importance of discovering new nematode species involves issues of science and economics. Scientific measures of importance are often broad in scope, deal with long time scales and are difficult to quantify in monetary terms, whereas economic measures of importance are generally limited in scope, refer to relatively short time scales and are readily quantifiable.

Our possible inability to quantify the immediate economic impact of discovering new nematode taxa does not indicate a lack of economic consequence, as recent calculations of the benefits of long-term "ecosystem services" have shown large dollar values. Nonetheless, the priority of discovering new taxa in different groups differs, depending on the relative weight given to the two ways of measuring importance, and this should be explicitly detailed. The plant-parasitic Tylenchida, insect-parasitic Rhabditida, and vertebrate-parasitic Strongylida and Ascaridida are examples of nematode orders that have clear economic importance (and of course these are also the best inventoried). In terms of ecosystem services and other long-term criteria, free-living nematodes from the other nematode orders have clear importance.

Sampling

When planning a nematode survey or inventory, protocols of sampling, extraction and specimen processing must result in specimens that are amenable to both morphologic and molecular analysis. Further, metadata collected at the time of sampling should accommodate the multiple uses of survey and inventory data.

The goal of a nematode survey or inventory is to capture information about the presence of nematode species occurring in one or more landscapes, ecosystems or geographic regions. Hence, the objectives and priorities of a survey or inventory determine the characteristics of an adopted sampling protocol, which ideally should represent an optimal mix of expertise, resources and time. Sampling details, such as substrate sampled, sample number, sample size, sample location and sample depth, warrant consideration when developing a protocol. However, the overriding issues of sampling pertain to sampling scale, pattern of sample collection and strategies of metadata collection.

Scale: The scalar dimensions of a survey or inventory relate to the breadth and depth of the sampling effort. Sampling breadth denotes the relative number of sites from which samples are collected, whereas sampling depth denotes the extent of taxonomic characterization per sampling site. A survey or inventory can rarely characterize a large number of taxa from a large number of sites. Apportioning resources between sampling breadth and depth entails a choice between an extensive or intensive sampling protocol.

An extensive protocol maximizes the number of sampling sites and characterizes a limited number of taxa per site. An intensive protocol limits the total number of sampled sites but maximizes the number of characterizations per site. The rationale for conducting a survey or inventory will determine whether an intensive or extensive approach is more appropriate.

Collection. The next consideration in developing a sampling protocol is the spatial and temporal pattern of sample collection. When the principal investigator directing the survey or inventory retains control of the entire sampling process, the patterns of nematode distribution should take precedence over project logistics in developing the collection protocol. However, if the survey or inventory relies on the contributions from many cooperators, who might collect samples at different times over a wide geographic area, then logistics inevitably becomes the driving consideration in sample collection.

A distributed collection pattern takes advantage of cooperator effort to collect samples based on broad sampling criteria. Such a strategy can generate new data at nominal expense and, for some purposes, results in information that satisfactorily addresses a pressing lack of knowledge about nematodes. More often, the rationale for conducting a nematode survey or inventory requires that samples be collected with greater rigor than is possible under a distributed plan.

Alternative strategies for the rigorous collection of samples include random, proportional, stratified and selected collection patterns. A random collection pattern assumes nematodes are uniformly distributed throughout sampled habitats. This assumption might be justified when nematode distributions are unknown or when a limited geographic area is being sampled.

A proportional pattern apportions the collection of samples based on representation of habitats within sampling sites. The number of samples collected from a particular habitat might be proportional to the geographic extent of the habitat within a landscape. Similarly, a stratified pattern weights the importance of habitats, and then, highly weighted habitats are more intensively sampled than low-weighted habitats. Under a stratified collection plan, all habitats are sampled to varying extents, and the relative importance of different habitats is determined by the rationale governing the survey or inventory. A selected pattern of sample collection restricts sampling to a single habitat or a small number of habitats within the sampled sites.

Metadata: When developing a sampling protocol, accommodation should be made to facilitate the future use and interpretation of survey or inventory data by the inclusion of associated (meta)data.

The kinds of metadata that might be collected include: methods of extraction, preservation of specimens, geographic coordinates, characteristics of the vegetation, selected soil physical and chemical properties, and information about soil biota other than nematodes (which might require coordination of the sampling effort with specialists in other organismal groups). The challenge posed by metadata is to identify it, collect it, and archive it. These activities should represent an integral part of the nematode survey or inventory, thereby conferring greater value on the information about nematode species occurrences.

Morphologic Characterization

A necessary first pass in any analysis will assign material to morphotypes. In many, and perhaps most, cases, material will not be assignable to named species. However, material can be identified to some taxonomic level (family, genus) and can subsequently be passed to appropriate experts and to laboratories where molecular analysis can be performed.

Using morphology without biochemical/molecular tools may be the most approapriate approach where the objective is initially to quickly focus on an estimate of trophic groups in an ecosystem. Such quick estimates of trophic groups might particularly be justifiable where there are adequate vouchers and collection site data (i.e., as is the case for many of the NSF Long Term Ecological Research sites), as this allows for future refined assessments.

All surveys include a step in which nematodes are sorted into morphologically discrete units. These units may vary in taxonomic content due to limitations of morphological characters for delimiting nematode taxa. Morphological units may actually consist of assemblages of cryptic species or ecotypic variants of a species. It is essential that biochemical and molecular-based approaches link the biochemical/molecular phenotype to these morphological units. Photographic morphometric analysis and the maintenance of voucher specimens will reinforce the linkage complementation between morphological and molecular approaches.

Use of Molecular Approaches

Extraordinary species diversity, the paucity of trained nematode taxonomists, and the difficulty of using morphologic characters for species-level identification challenge nematode taxonomy. New approaches are needed to aid species identifications within the context of a classical morphological system. Without the ability to easily and efficiently discriminate among species, nematodes will remain excluded from many areas of biological research.

Taxonomic surveys of soil nematodes could be made much more efficient by the development of molecular techniques for species-level identification of individuals. The exact nature of the molecular analysis may vary from study to study. However, we identify here the major criteria with which to evaluate candidate systems. Whatever system is chosen, the principal investigator will have to address several issues outlined below and justify their choice with respect to the goals and scale of the proposed study and the potential for future use of their data by other researchers.

Species-level resolution or diagnosabilty. The system must be able to discriminate species. The molecules chosen must evolve rapidly enough to display species-level differences; however, high levels of intraspecific variation will complicate species definitions. Phylum-wide approach. A molecular approach that can be universally applied will be more useful for taxonomically diverse surveys than group-specific approaches.

Character State Data: The data should be amenable to phylogenetic analysis. Ideally, such data should include clearly homologous characters which are useful in the broadest possible context. Character state data is important for the incorporation of cladistic species concepts. In addition, such data will contribute to the overall understanding of nematode relationships.

Identification from All Species Forms: In surveys and inventories, each species will usually be represented by different and/or multiple sexes, different forms of each sex (in more complex life cycles), eggs, and all juvenile stages including dauer juveniles. An accurate molecular method must identify all recovered stages of development.

Single Nematode Identification: The method employed must work on individual nematodes allowing for evaluation of intra-specific variation. Without prior knowledge of species-level identification, it is not possible to pool nematode samples for molecular analysis.

Practical for Large-scale Species-level Taxonomy: The method should be practical, routine, and possess the potential for future automation. The far-reaching goal of a species-level taxonomy for nematodes is enormous and will ultimately require large-scale data collection and analysis.

Amenable to Database/analysis: Methods which facilitate the collection of unambiguous and non-subjective communication of taxonomic characters are favourable. Data must be in a form that is useful to all nematode taxonomists.

Comparable Datasets: The availability of comparable datasets for homologous molecules in other taxa will provide for useful comparisons.

Voucher Specimens

The value of a biotic survey or inventory is measured by conservation and accessibility of its data. This includes data consistent with the immediate objectives, and data with added on value for ongoing and future complementary investigations. Specifically, when collecting nematodes, other nontarget taxa incidentally may be collected.

Collection, curation and exchange of nontarget groups, including specimens and molecules, is particularly desirable if it maximizes data from limited collection resources. Such cooperative exchange is strongly encouraged. Conservation of data must include preparation and deposition of morphological voucher specimens by methods which maximize long-term preservation, curation, and availability in at least one major institutional collection (as recognized by the discipline).

When surveys or inventories include molecular or biochemical data, provision should be made for appropriate linkage (i.e., databases) to morphological vouchers represented by either the actual specimen from which DNA is extracted, especially when that extraction is designed to minimize loss of morphological data, or additional specimens from the target population. When combined with morphological voucher specimens, photographs/drawings/and morphometrics derived from specimens prior to microdissection for molecular characters are important for linkage between morphological and molecular data. A particular survey or inventory must consider the potential and value for conserving a range of additional data including frozen tissue for molecular extraction, as well as frozen-living and active cultures.

Alternate Strategies for Nematode Control Towards Sustainable Agriculture

A large proportion of the food production in many tropical and subtropical countries comes from smallholders farming relatively small parcels of land. Crops may be grown together in mixed cultivations often following a system that has been long established and generally successful. In the era of "progress" signalled by the development of monoculture with new varieties, crop protection aids and inorganic fertilizers, it is regrettable that in hindsight the knowledge of yesterday contained in the practices and methods of these smallholders has not been given more attention.

The trend towards urbanization has generated the need for regular supplies of food to be produced from farms and smallholdings close to cities, or with reasonable access to a road network. Intensification of food-crop production to meet this consumer demand has caused various problems for agronomists, soil scientists and crop protectionists, particularly if the land is in continuous cultivation. The traditional system of agriculture relied on the use of long periods of fallow between crops. During this time the natural vegetation regenerated, soil fertility was restored and populations of pests and diseases declined. Traditional farming systems are usually of mixed crops (even including trees or woody perennials) that provide an ecosystem in which total crop loss due to pests and diseases is rare.

Nevertheless, subsistence farmers will usually grow enough to cover the possibility of losing up to 40 percent of production before or after harvest. The problems of a rapidly increasing human population for land use has led to the abandonment of traditional systems, and these are now practised only in sparsely populated countries. Crop losses due to pests and diseases appear to be more frequent as improved crop varieties are grown in monoculture on the same land with the aid of pesticides and inorganic fertilizers. In many instances, crop damage has increased as a result of the inadvertent introduction of alien pests on crop plants. The distribution of several nematode species has been increased in this way, e.g. *Radopholus similis* and *Pratylenchus* spp. on bananas, *Tylenchulus semipenetrans* on citrus and *Globodera* spp. on potatoes.

In considering potential strategies for controlling nematodes there are certain practices that can be implemented once the life history and the behaviour of the pest is known. Plant-parasitic nematodes occur in greatest abundance in the rooting zones of crops, mostly in the top 30 cm of soil; nematodes may also penetrate to deeper soil layers particularly in unfavourable conditions only to return to the surface layers when crop growth resumes. The free-living stages of some species survive the unfavourable season through physiological or structural adaptation, or survive for several seasons as eggs, as in the case of the potato-cyst nematode.

Nematodes are usually at their most vulnerable at the time when they are actively searching for host roots and when surviving unfavourable growing seasons. The survival of free-living stages in the absence of hosts can be longer than originally thought, and fallow periods of less than 12 months may be insufficient to control those species that reproduce rapidly. The main abiotic factors that influence free-living nematodes and eggs are heat, desiccation and anaerobis. Plant-parasitic nematodes will not survive long periods at temperatures above 40°C. These temperatures may be reached on bare soil in some countries during the hot season, but heat penetration may not extend below 10 cm. The technique of increasing soil temperatures by solarization can be of value in controlling nematodes, but this is only possible where there are long periods of uninterrupted sunshine. Solarization is also practised for controlling soilborne fungal pathogens. Nematodes are aquatic animals that inhabit the films of water that surround soil particles.

As soils dry, the volume of soil water decreases and the ability of nematodes to move is impaired, eventually free-living nematodes may die as the soil moisture decreases. The extent to which species have adaptations to survive desiccation is still a matter for further study. By regulating the metabolism in times of stress some species are capable of anhydrobiotic survival. The contrivance of bare fallows that deprive nematodes of a host may not always be a worthwhile control strategy especially as fallowing in hot, dry areas can deplete soil organic matter and the levels of beneficial soil biota. In such situations for specific nematode pests, a better alternative might be cover crops with non-host leguminous trees or shrubs such as *Crotalaria* spp. The control of nematodes by flooding has been advocated in certain locations. When flooded all the soil pore spaces are filled and the oxygen supply for the soil microflora and fauna becomes limiting.

Many plant-parasitic nematodes are intolerant of oxygen starvation and soon die. Similar effects are observed when nematodes are stored in deep water in the laboratory, but often they can be revived by aeration. To be effective in the field therefore, the anaerobiosis has to be of sufficient duration to kill the nematodes. In Côte d'Ivoire banana fields are flooded for at least five weeks. Flooding is not a technique that can be widely used for nematode control and whether it is equally efficacious against all plant-parasitic species is not known.

The development of strategies for nematode control requires a thorough understanding of the growth of the host plant and the biology of the nematode. In some climates it is possible to avoid (or contain) damage by planting the crop at a time when nematode invasion will be slow, or when plant growth will be faster than that of the parasite.

Damage to transplanted short-cycle crops can be limited if seedlings are produced in nematode-free nurseries or in soil-less potting media. In general, protection from invasion during the early growth of a crop will enable the plant to produce a yield acceptable T most farmers even if it is planted in nematode-infested soil, particularly if the crop receives adequate supplies of water and nutrients. This will not prevent the development of large nematode populations on the root systems and in the soil that will influence the subsequent crop. Brown (1987) showed that treating wheat with low nematicide doses at seeding gave a substantial yield improvement, resulting from protection from the invasion of cereal-cyst nematodes during the establishment of the crop.

The concept of growing companion plants with nematicidal root exudates alongside susceptible crops is attractive, but the actual benefits may be questionable or at least difficult to quantify. Species of *Tagetes, Crotalaria* and castor bean (*Ricinus communis*) have been widely used. Where *Tagetes* was grown as a companion to tomato in a rotation trial in the United Republic of Tanzania, yields were significantly less than with other treatments, indicating that such a cropping practice was not likely to be adopted by farmers.

The use of organic mulches for managing nematodes has been widely studied and there are various examples of plant debris having beneficial effects on plant growth. In small garden plots that receive heavy mulches of organic waste, nematodes are unlikely to be a major problem. Subsistence gardens are therefore well suited to the needs of householders, who, as Hussey (1990) reported, will usually grow sufficient quantities to meet immediate needs and allow for crop losses due to pests and diseases. Regular mulching is realistic for subsistence gardeners but becomes less feasible if the area of cultivation is large. To be effective, the recommended mulches may have to be used at application rates of 5 to 10 tonnes per hectare. The acquisition and distribution of such volumes is not likely to be practised by farmers growing crops on a large scale.

Future Developments

If chemical treatments are unavailable or inappropriate, the management of plant-parasitic nematodes will have to depend on alternative strategies.

Those farmers in the Near East region who routinely use methyl bromide or non-fumigant nematicides may in future have to seek alternative treatments. To help them do so, government-supported research will be necessary, particularly for control strategies not requiring the regular purchase of proprietary products. This type of research is unlikely to be funded by manufacturing companies.

New research will require multidisciplinary collaboration to enable the integration of traditional practices with new ideas created by the scientists. Through more research on nematode biology and the environment, nematologists must consider several possibilities:

- How nematode life cycles may be interrupted.
- How microbial activity at the root-soil interface, which could lessen nematode invasion, can be promoted.
- Identification of compounds that interfere with reception of stimuli or repel nematodes.
- The selection or breeding of resistant varieties, or varieties that tolerate nematode infestation, and that can produce acceptable yields.

In conducting field experiments that evaluate the effect of cropping systems on nematode populations, scientists should plan to suit the needs of the farmer. Many research station trials are designed to produce results for ready analysis and presentation in journals; in consequence, treatments may be selected to suit the needs of the scientist. Most farmers grow short-cycle, annual and perennial crops in close proximity. Such mixed cropping practices are rarely undertaken in field experimentation by nematologists.

Any new cultural, microbial or management approaches that are developed must be within the capabilities of the farmer and meet the necessary environmental and economic requirements. New strategies will be considered as successful if they lower nematode damage thresholds. The farmer will be satisfied if these treatments are reliable, practicable and economically justified; his customer, the consumer, will be content if the product has the desired quality, contains no toxic residues, and is fairly priced.

Chapter 11

Nematodes and Plant Damage

Function of Roots

- absorb water
- absorb nutrient ions
- storage
- anchorage
- dynamic mining function, regeneration.

Terrestrial plants are estimated to have annual production of 120 billion tons of biomass -5% minerals = 6 billion tons of minerals mined from the soil each year. A plant may transpire its own weight in water in a day. The integrity of Casparian strip (waxy layer around endodermis cells) is important-nutrients, and sometimes water, are taken up against a gradient. Secondary roots and endoparasites are disruptive to the Casparian strip.

Components of Damage

Demand

Total energy consumption during the lifecycle of a female root-knot nematode (*Meloidogyne* spp.) is 1 calorie.

The total biomass of a female root-knot nematode is 200 μg, including the egg mass (Melakerberhan and Ferris). For say 100,000 nematodes in a root system, the total nematode biomass is 20 g! Allowing for 50% production efficiency, total material extracted from the plant would be 40 g. So, the demand effect on the plant may be minimal unless plant is very stressed and resources are limited.

An adult *Heterodera schachtii* consumes 11 nL/day of cell content. So, it would take 1,000,000 such females to remove 11 ml of cell content in a day.

Mechanical Disturbance

a. *Penetration of cells*-relative to length of stylet. Damage will depend on types of cells affected-storage tissues, cortex, or functional vascular.

b. *Migration through tissues*-intercellular and intracellular requiring dissolution of cell walls, middle lamellae. Suggests cellulase and pectinase enzymes-spongy tissues, sloughing, e.g. damage caused by *Pratylenchus* and *Ditylenchus*. Allows ingress of other organisms. Root-knot (*Meloidogyne* spp.) and cyst (*Heterodera* spp.) produce endogluconase (cellulase) enzymes and pectate lyase which are presumable involved in the passage through plant tissues.

c. *Leakage from damaged tissues*-it is estimated that up to 20% or more of photosynthate partitioned to roots may leak into rhizosphere soil without root damage. "Root exudation" nurturing organisms in rhizosphere-presumably to plant benefit-but speculate that selection has optimized the costs and benefits. Enhancing root leakage through nematode damage must reduce plant productivity.

Physiological Disturbance

Nematode secretions-associated with establishment and maintenance of feeding sites. Effects increase with sedentary endoparasitism. Secretions from the nematode digestive glands may polymerize into a feeding tube inside the cell. The feeding tube remains associated with the stylet during ingestion. When the stylet is withdrawn the opening in the cell wall is sealed with an electron-dense feeding plug.

- Cell wall permeability and surface enlargement-transfer cells, cell wall thickening.
- Increase in cytoplasm density, metabolic activity.
- Hypertrophy-cell enlargement, 500 to 1000x increase in volume (*Meloidogyne*)-nuclear division without cytoplasmic division (karyokinesis without cytokinesis).

But different strategies in different genera-eg cell wall dissolution in *Heterodera*. Multinucleate cells, nuclei larger.

- Hyperplasia-tissue enlargement-mitotic activity-galling, root tip galls etc.
- Disruption of normal meristem activity-*Xiphinema, Trichodorus, Hemicycliophora* (different taxonomic groups). Damage to meristems.
- Physiological control of carbohydrate partitioning-Metabolic sink-McClure, $^{14}CO_2$ studies.

Physiological effects:-

- increased root respiration (more mass),
- increased growth respiration-repair,
- mobilization of defence mechanisms.
- increased levels of plant hormones (indole acetic acid, cytokinins) in galled tissue, but source uncertain.

Whole-plant effects-Disturbance of the biochemical network. Wallace (1987) points to the complexity of the biochemical pathways:

Photosynthesis divided into two basic phases-a light phase when light energy is converted into chemical energy, and a synthetic phase in which carbohydrates are formed in a series of reactions accelerated by light. Photosynthesis involves a chain of metabolic events cross-linked to other physiological processes, so disruption of one may have effects throughout system. For example, Bird suggested that photosynthesis is reduced in tomato by *Meloidogyne javanica* by inhibiting production of cytokinins and gibberellins in roots, and/or by increased stomatal resistance due to water stress.

Fatemy et al. indicate that the response of potato to *Globodera rostochiensis* is due to stomatal closure through water stress; the result is reduced photosynthesis. However, generally the mechanisms by which root-infecting pathogens, including nematodes, affect physiological processes have been insufficiently studied.

Plant as an Integrator-Metabolic pool concept-plant as an integrator-concepts of demand and damage. Melakeberhan and Ferris characterized five effects of root-knot nematode infection in grape while exploring the impact in an energy partitioning and flow model:

- Reduction (disruption) of water uptake. Seinhorst, however, asserted that there is little evidence of reduction of water uptake in response to nematodes. He measured daily water uptake on a water usage basis-rate of water loss from pots minus increase in dry wt minus evaporation from surface. The rate was a linear function of total dry wt. So, he argued that rate of use per g tissue is constant. However, root damage could result in lower water uptake, and final dry weight could be a function of the rate of water uptake.
- Reduction in rate of photosynthesis.
- Reduced leaf expansion and total photosynthesis.
- Alteration of partitioning of photosynthate-change in root/shoot ratio.
- Increased leakage-direct effect and affect on other pathogens energy supply (Garrett-inoculum potential as a function of the abundance of infective units and the energy resources available to them).

Molecular Events

Plant responses to nematode indection (from various plant and nematode systems):

- Pathogenesis-related proteins (similar to those after fungal or viral infections) are induced.
- Non-local induction of proteinase inhibitor proteins occurs.
- Local and systemic induction of a gene encoding catalase occurs. Catalase may protect the plant from oxidative stress associated with nematode feeding.
- Expression of a gene encoding extensins, which are hydroxyproline-rich glycoproteins that are components of plant cell walls.

Changes in gene expression at feeding sites is probably localized in a few cells. Identifying the small amount of product is technically difficult. Use of reporter genes such as GUS which can be attached to promoter regions of other genes are among useful new indicators of localized gene activity.

Evolutionary Events

The genome of plant-feeding nematodes of the Order Tylenchida includes genes that encode for endoglucanases. Endogluconases are cellulases, a family of enzymes formerly thought to be restricted to prokaryotes. Other plant-cell wall digesters such as termites and ruminants use symbiotic and commensal bacteria to achieve dissolve cellulose. The presence of these and other genes suggests that horizontal or lateral gene transfer has occurred between bacteria and nematodes.

Proof of pathogenicity

Koch, Pasteur-the germ theory-required rules of proof. Mountain provided guidelines for obligate parasite nematodes:

- establish association-ecological phase
- establish parasitic capabilities-biological phase
- establish host-parasite relations-etiological phase.

Predisposition

Interaction with other organisms

- other nematodes
- fungi
- bacteria
- viruses
- abiotic/physical stresses.

Note-term "interaction" is loosely used-implies that effect in combination is different than sum of individual effects-not additivity. Three descriptions of the result of combinations of organisms:-synergistic,-suppressive,-no interaction.

Mechanisms of Interactions

Synergistic

- Reduced tolerance-multiple stress
- Vectoring-Longidoridae and Trichodoridae
- Change in substrate-fungi?
- Routes of ingress-fungi, bacteria
- Leakage-energy source-fungi, bacteria.

Suppressive

- Induced resistance-any examples?
- Reduction of stress-mycorrhizae
- Biological antagonists-but not an interaction of two pathogens
- Reduced substrate availability.

Importance of Nematodes in World Agriculture

See Sasser and Freckman in Vistas on Nematology. Questionnaires returned by 371 nematologists worldwide. Consider inherent biases in data of this kind. Consider means and variance in crop loss data. -almost nobody had 10% yield loss. -include management costs as part of the loss. Ranking of important genera and relative weight:.

Genus	*Relative Damage Importance*
Meloidogyne	1.00
Pratylenchus	0.57
Heterodera	0.44
Ditylenchus	0.18
Globodera	0.19
Tylenchulus	0.17
Xiphinema	0.15
Radopholus	0.12
Rotylenchulus	0.10
Helicotylenchus	0.09

Note-considerable variation by region, so questionnaire data biased by number of respondents per region.

International Survey of Crop Losses due to Nematodes

Life-sustaining Crops	***Annual Loss (%)***	***Economically-important Crops***	***Annual Loss (%)***
Banana	19.7	Cacao	10.5
Barley	6.3	Citrus	4.2
Cassava	8.4	Coffee	15.0
Chickpea	13.7	Cotton	10.7
Coconut	17.1	Cowpea	15.1

Corn	10.2	Eggplant	16.9
Field bean	10.9	Forages	8.2
Millet	11.8	Grape	12.5
Oat	4.2	Guava	10.8
Peanut	12.0	Melons	13.8
Pigeon pea	13.2	Misc. other	17.3
Potato	12.2	Okra	20.4
Rice	10.0	Ornamentals	11.1
Rye	3.3	Papaya	15.1
Sorghum	6.9	Pepper	12.2
Soybean	10.6	Pineapple	14.9
Sugar beet	10.9	Tea	8.2
Sugar cane	15.3	Tobacco	14.7
Sweet potato	10.2	Tomato	20.6
Wheat	7.0	Yam	17.7
Average	10.7%	Average	14.0%
Overall Average 12.3%			

Information based on a worldwide survey with 371 responses.

Source: Sasser, J.N., Freckman, D.W., 1987. A world perspective on Nematology: the role of the society. Pp 7-14 in J.A. Veech and D.W. Dickson (eds) Vistas on Nematology. Society of Nematologists, Hyattsville, Maryland. 509p.

Crop losses-measurement and estimates

- -Problem of reference yield-how do you isolate one pest effect?
- -Need for holistic approach.
- -How to remove one stress from system without changing system.

Vectors of Plant Viruses

Rev.

- Viruses are categorized into over 30 distinct groups which are classified as families and genera.
- They include viruses which consist of double-stranded DNA, single-stranded DNA, double-stranded RNA and single-stranded RNA
- Two single-stranded RNA virus genera, Nepovirus (NEPO) and Tobravirus (TOBRA), have nematode vectors.

- Nepoviruses are in the Comoviridae family while Tobraviruses are not yet assigned to a family.

Nepoviruses (Nematode-transmitted Polyhedral viruses) and Tobraviruses (Tobacco Rattle viruses) have nematode vectors.

Nepoviruses are isometric (polyhedral) particles around 30 nm in diameter (1 nm=10^{-9} m, 1 µm=10^{-6} m). The only known nematode vectors are in the genera *Xiphinema* and *Longidorus*.

Tobraviruses are straight tubular particles with two size ranges, 180-210 nm and 45-115 nm. *Trichodorus* and *Paratrichodorus* are vectors.

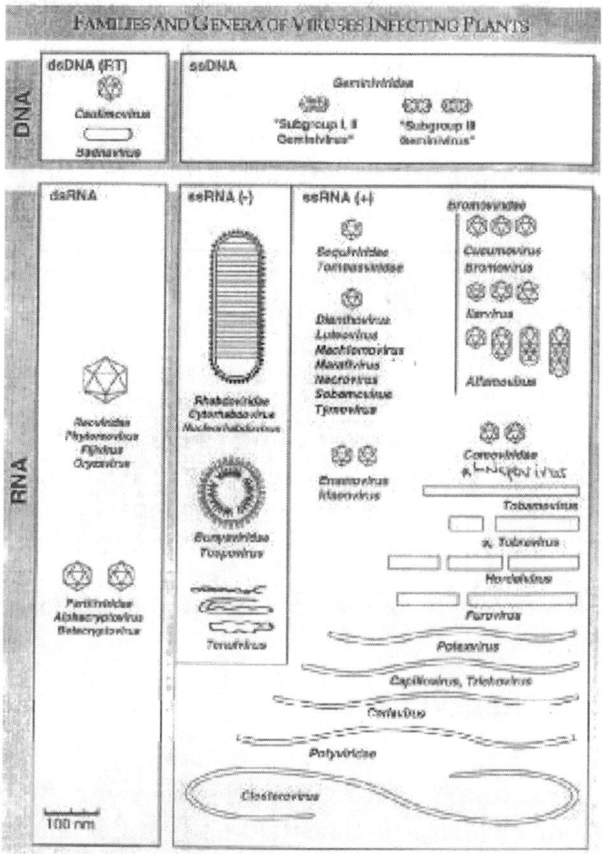

The demonstration by Hewitt *et al.* (1958) that certain nematodes are vectors of plant viruses initiated research in Nematology and Virology that resulted in understanding of the transmission and etiology of an important group of soilborne plant virus diseases.

Vector Specificity

11 spp of *Xiphinema* transmit 13 NEPO viruses

11 spp of *Longidorus* transmit 10 NEPO viruses

14 spp of *Trichodorus* transmit various strains of two TOBRA viruses:

tobacco rattle and pea early browning.

Lamberti (1987) suggests that since several trichodorid spp transmit the same virus, and that both viruses are transmitted by the same nematode, vector specificity is less developed in trichodorids than in longidorids.

Transmission Characteristics

Trichodorids may retain the virus for up to a year.

Acquisition time may be less than an hour to several days, depending on the feeding characteristics of the nematode.

Retention sites: *Longidorus*-odontostyle area:

Xiphinema-odontophore and esophagus region

Trichodorus-onchiostyle and esophagus.

Selectively and specifically adsorbed at retention site-indicating a specific association between protein coat of virus and cuticular surface.

Dissociation with cuticle probably occurs as glandular secretions pass forward into the plant cell.

- Virus does not pass through the egg stage.
- Virus does not replicate in the nematode.
- Virus is not retained through a molt.

In *Xiphinema* and *Trichodorus*, although the lining of the esophagus is not shed at a molt, it undergoes structural changes and virus particles may pass into the intestine. In *Longidorus* the stoma, odontostyle and guiding sheath are shed.

Nepoviruses

X. americanum (sensu lato) transmits tobacco ringspot and tomato ringspot.

The virus is acquired within 24 hours of the initiation of feeding.

Virus particles are transmitted by both adults and juveniles.

X. americanum transmits the tomato ringspot virus strains:

- peach rosette mosaic virus
- peach yellow bud mosaic virus
- cherry rasp leaf virus
- grapevine yellowvein virus

Particles are carried in the esophagus lumen.

Lamberti et al. consider that *X. americanum* is not a vector of tomato ringspot virus, although it is widely reported to be. They contend that the reports represent a mis-identification especially in California where they feel that the vector is *X. californicum*. However, Griesbach and Maggenti (1989) found a wide range of vectoring capabilities of tomato ringspot viruses with California populations of *X. americanum* sensu lato.

On the basis of considerable overlap of morphological and morphometric characters among putative species, they synonymized *X. americanum* and *X. californicum*. That synonymy was later rejected. The situation is not yet clearly resolved; studies on developmental biology and the application of molecular techniques are providing further information on the diversity within the whole *X. americanum* group.

Lamberti suggests that the same mis-identification occurs with the vector of Cherry Rasp Leaf virus in California.

There is a need for the application of molecular biology techniques in this problem of diagnostics.

X. californicum transmits prune brown line and *Prunus* stem pitting strains of tomato ringspot virus readily and the cherry leaf mottle strain rarely.

X. diversicaudatum transmits *Arabis* mosaic virus to variety of crops.

Adults retain the virus for at least 8 months.

X. index-grapevine fanleaf virus can be acquired in 5 to 15 minutes, persists up to 9 months when nematode not feeding.

Longidorus elongatus transmits raspberry ringspot and tomato blackring viruses.

Docking Disorder

Longidorus leptocephalus, L. attenuatus, L. elongatus, Trichodorus spp and viruses are involved in Docking disorder of sugarbeets (named for Docking region in south of England). Stunted growth in spring due to nematodes, virus symptoms in foliage. Effects are most pronounced in spring, heavy rainfall in May seems to increase the problem. In July the affected plants start to grow again and may achieve almost normal foliage, but a much reduced tap root. The severity of Docking disorder varies from year to year with the climate.

Management by rotation is difficult as several nematode species are involved, each with a differing host range. Also host ranges of these nematodes are not completely known. 1,3-D nematicide applied in the plant row prior to planting reduces the nematode populations and increases yield, but only in well-drained alkaline soils. The treatment is not always effective at economically feasible rates.

Grapevine Fanleaf Virus

Grapevine Fanleaf Virus Degeneration is the oldest known disease of *Vitis vinifera*.

Records in Europe date back 200 years and specimens in herbaria displaying symptoms predate the introduction of American rootstocks. Grapevine Fanleaf Virus Degeneration is the oldest known disease of *V. vinifera*.

The observation that *X. index* transmits grapevine fanleaf virus (GFLV) was the first record of virus transmission by a nematode.

The virus is lost during the molt of the cuticle between life stages. The virus is not passed through the egg stage. Consequently, the virus is re-acquired by feeding of each vermiform life stage of the nematode. GFLV causes reduced vigor, lack of fruit set, and reduced yield of grapevines. Of great importance in the design of control programs for the *X. index*/GFLV problem is that portions of the grape root system can remain alive and serve as a reservoir for virus and nematode for at least 5 years after vine trunks have been removed.

The grapevine fanleaf virus is a Nepovirus with isometric particles of 30nm diameter. It is transmitted by *X. index* and *X. italiae*, with *X. index* the more efficient vector.

The virus can be acquired in less than 15 minutes of feeding on an infected plant and is retained up to eight months in the absence of feeding.

The retention site for GFLV in *X. index* is the cuticular lining of the esophagus, including the area surrounding the odontophore. As with other viruses vectored by longidorid and trichodorid nematodes, it is selectively and specifically adsorbed at the retention site, indicating a specific association between protein coat of virus and cuticular surface. Dissociation with cuticle probably occurs as glandular secretions pass forward into the plant cell.

The virus does not pass through the egg stage. It is not retained through a molt when the esophageal lining and odontostyle are also shed. In *Xiphinem*a, although the lining of the esophagus is not shed at a molt, it undergoes structural changes and virus particles may pass into the intestine.

The virus does not replicate in the nematode vector.

Detection: From a management standpoint, it is important to determine whether *X. index* is present in a site intended for a vineyard, and whether the nematodes are viruliferous.

Virus particles can be detected in extracts from single nematodes by immunosorbent electron microscopy, however, as indicated by Esmenjaud et al (1993) the procedure is complex and not readily adapted to routine assays.

Enzyme-linked immunosorbent assay (ELISA) methods are routinely used to determine GVFL in plant tissue. When applied to nematodes, the techniques appear sufficiently sensitive to detect the virus if more than ten viruliferous individuals are used, but are unreliable for single individuals.

Symptoms: Martelli and Savino (1988) describe three distinct syndromes of symptoms of grapevine fanleaf degeneration:

1. Malformations: Leaves are variously distorted, asymmetrical, with wide petiole sinuses and abnormal vein arrangement that results in the appearance of an open fan. Leaf symptoms may also include chlorotic mottling. Foliar symptoms are most evident in the spring and through the vegetative growth period. Shoots show abnormal branching, with double nodes and short internodes, while forming angles at internodes. Bunches are small with poor fruit set and shot berries. They ripen irregularly.

2. Yellow mosaic: Bright yellow discoloration of leaves, shoots, tendrils and inflorescences in the spring. Foliage and clusters are not malformed, but clusters are small.
3. Veinbanding: Yellow flecking along main veins and associated interveinal areas of mature leaves appearing mid to late summer. Leaves are not malformed, but fruit set is poor and yield is low.

Several closely related nepoviruses found in grapevines are apparently related to GFLV but are serologically different. They are sap-transmissible, and where vectors have been definitively established they are species of *Xiphinema* and *Longidorus*. They include arabis mosaic virus, raspberry ringspot virus, tomato blackring virus, grapevine chrome mosaic virus, Strawberry latent ringspot virus, artichoke Italian latent virus, and grapevine Bulgarian latent virus.

One or more of ten species of *Xiphinema* are present in all major grape-growing regions of the world. They are *X. algeriense*, *X. americanum*, *X. brevicolle*, *X. diversicaudatum*, *X. index*, *X. italiae*, *X. mediterraneum*, *X. pachtaicum*, *X. turcicum*, and *X. vuittenezi*. With *X. index* a notable exception, most species of *Xiphinema* have a wide host range and are adapted to a wide range of soil textures.

Xiphinema americanum is actually considered a group of some 39 species, which include *X. brevicolle* and *X. pachtaicum*. Molecular techniques, biological evaluations and virus transmission studies are continuing to underscore the variability and similarities within the *X. americanum* group. These studies are adding to the body of knowledge that will be required for an eventual taxonomic revision. The group also includes *X. californicum* Lamberti and Bleve-Zacheo, which is probably associated with grape in California.

Tomato Ringspot Virus in Grapevines

Tomato Ringspot Virus Degeneration is also called grape yellow vein disease, tomato ringspot disease, and little berry disease. It is endemic in the northeastern U.S. and Canada, and occurs less frequently in California. The nepovirus is transmitted by *X. americanum*, *X. californicum*, and *X. rivesi*. Particles are carried in the esophagus lumen adhering to the cuticular lining.

Since there is considerable variability in the species complex that is loosely-termed *X. americanum*, Lamberti and Bleve-Zacheo (1979) contend that reports of *X. americanum* as a vector of TRSV represent a misidentification, or predate taxonomic revisions. They consider that *X. californicum* is the vector of TRSV in California. However, Griesbach and Maggenti (1989) found a wide range of vectoring capabilities of TRSV strains with California populations of *X. americanum sensu lato*.

Symptoms of Tomato Ringspot Virus Degeneration differ geographically, with the disease more severe in colder regions. Leaves are mottled, with oak leaf patterns, new growth is weak and buds are susceptible to cold damage. Internodes are shortened, leaf area is reduced, and clusters are small. The grape yellow vein symptom, common in infections with this virus in California, consists of yellow flecking along the veins and associated lamina. There is reduced fruit set and slow decline of the vine.

Peach Rosette Mosaic Virus in Grapevines

Peach Rosette Mosaic Virus Decline is a disease of *V. labrusca* L. grapevines that occurs only in Michigan in the U.S. The virus also causes a disease in peach, as indicated by the name. This nepovirus is a strain of TRSV transmitted by *X. americanum* and *Longidorus diadecturus* with the characteristics of other nepoviruses transmitted by these genera. The virus may be endemic in vineyard weeds, which complicates exclusionary measures. Symptoms of Peach Rosette Mosaic Virus Decline include puckered leaves with flattened basal sinuses, sparse clusters, and crooked shoots. The vines are susceptible to winter injury.

Management Guidelines for GFLV-Pre-plant

Avoidance

The wide host ranges of ectoparasitic nematodes make pest avoidance difficult to achieve. One exception is *X. index* and GFLV. The use of virus-free and nematode-free rootings provide the best method of controlling the complex and certification programs have reduced its dissemination through vegetative propagation. The wide host range and current abundance of *X. americanum* in U.S. soils is an indication that quarantine of the nematode is not possible, but the certified virus-free nursery stock program continues to have merit. On-farm nurseries should be maintained nematode-free and located away from old vineyard sites.

Cultural Controls

By their nature, ectoparasitic nematodes can survive in soil or persist on weed hosts for long periods of time. Fallowing will only reduce populations, not remove them. The persistence of old grape roots in replant sites is a problem. New grape roots tend to follow the channels in the soil left by old roots. GFLV can be moved from these reservoirs to healthy planting stock through exploratory feeding probes by *X. index*. It may be difficult to develop 'immune' rootstocks on which the nematode cannot reproduce and never even attempts to feed. Replanting of a virus-infected vineyard with a rootstock that is resistant to *X. index* within 6-7 years after removing a virus infected vineyard will not be successful in the long term unless the nematode does not attempt to feed. The site will probably not be safe until 1-2 years after the last old root is dead. Such time intervals may not be economically feasible. Resistance to GFLV will be an important solution to the problem.

Old roots are most easily killed before the old vine is removed by using systemic herbicides. Once the vine is removed the only reliable method of killing roots in the top 2 m of soil is by soil fumigation. Soil ripping to that depth on 30 or 60 cm centres may increase the speed of root death, but this tactic has not been field tested. Heat is a useful root killing agent but we are unable to deliver the required heat using current methods. Soil flooding for months will not kill old grape roots or nematodes. Selected rotation crops will reduce population levels of certain nematode species over time. This will have little value for endoparasites but should be tested with ectoparasites.

Certain *Sorghum bicolor* L. x *Sorghum sudanense* Stapf hybrids grown in summer have negative effects on ectoparasites but the number of years required to eliminate a nematode problem is unknown. Addition of manure, composts and ammonia does not kill old roots, but there are various reports of the benefits of ammonia as a nematode control agent.

Chemical Controls

Soil fumigation with methyl bromide or 1,3-Dichloropropene is effective for killing old roots 1.5 and 2 m deep in soil. Such treatments can also give 99.9% reduction of all nematode species in the top 1.5 to 2 m of soil when properly applied. Two to six years after such treatments, the nematodes do return unless resistant rootstocks are replanted. The use of high rates of 1,3-dichloropropene was halted in the state of California in spring 1990. Methyl bromide use is to be phased out by the year 2000, and adequate replacement fumigants must be environmentally safe. Environmental problems with nematicides have also occurred elsewhere.

The future of all general biocides as a fumigation replacement is unclear. An approach we have taken is the integration of "softer" root killing agents or tools coupled with "softer" soil treatments which reduce soil-dwelling population levels. The practice of removing a vineyard and replanting within one year is not a current option in vineyard management where nematode problems occur. There will be much discovery in this area in the decades ahead. One treatment strategy worthy of study involves use of root killing and soil cleansing treatments followed by one or more years of non-host crops.

Management Guidelines for GVFLV-Established Vineyards

Avoidance

Movement of contaminated grape harvesting equipment and tractors from a site infected with *X. index* should be avoided.

Biological Control

Refer to the previous chapter on this subject however, the addition of biological agents to soil has been inconsistent and usually ineffective. It may be possible, however, with the advent of drip irrigation systems, to apply microbe-produced toxins directly to the soil in irrigation water. Zoosporic fungal parasites have been isolated from *X. rivesi* and *X. americanum*. Such fungi require free water for their movement through soil. Their occurrence suggests research to investigate water management protocols that will enhance fungal parasitism of dagger nematodes.

Cultural Controls

By reducing vine stress through more frequent irrigation the damage caused by nematodes can be reduced.

The use of grassy cover crops in vineyards infested with *X. index* should be studied. However, legume cover crops in vineyards should be monitored to assess population levels of *Criconemoides xenoplax,* which may increase. Populations of *X. americanum* will also increase on most cover crop selections. The difficulty in choice of cover crops is that the host range of most ectoparasites is quite broad.

Nematode species with long bodies tend to be in shallow and non-disturbed sites of a vineyard, so placement of any treatment is important. Tillage and soil disturbance can reduce population levels short periods of time, but root surface area is also reduced.

Drip-irrigation-applied fertilizers that release ammonia may reduce population levels of ectoparasitic nematodes when applied repeatedly. Fifteen kg/ha of nitrogen in urea salt when re-applied three to five times at 30 to 45 day intervals can reduce population levels of most ectoparasites by half. More field testing of these strategies is necessary. Since grapevines do not have a high nitrogen requirement and some vineyards already have excess nitrogen, this technique should be tested and implemented with caution.

Chemical Controls

Organophosphate and carbamate nematicides are lethal to ectoparasitic nematodes when used as single treatments at high rates via drip irrigation. Treatments with currently-available commercial nematicides only reduce populations about 50% for 6 to 8 months after treatment. Multiple treatments with low rates of phenamiphos (1 kg/ha) are ineffective against ectoparasites. When multiple treatments are used against endoparasites for several years, population levels of ectoparasites such as *X. americanum* can be observed to increase above the non-treated population levels. In the case of ectoparasites, the value of systemic nematicides will be minimal unless the toxicant is available during feeding or leaks out into the rhizosphere.

Future Research

Except for *X. index,* ectoparasitic nematodes have not received the same research attention as endoparasites. Long-term experiments on the damage potential and management of ectoparasitic nematodes in vineyards are needed.

The apparent increase in incidence of *X. index* associated with reduced use of conventional soil fumigants is important and should be monitored. Other problems with ectoparasitic nematodes may similarly emerge. New pre-plant control strategies will be needed in the near future. Resistance and tolerance of rootstocks will continue to be important areas of study.

Tobra Viruses

Trichodorus viruliferous-transmits tobacco rattle and pea early browning virus. It is sometomes involved in Docking disorder in sugarbeets in England. *Trichodorus similis*-transmits tobacco rattle virus. At least 14 species of Trichodoridae have been demonstrated to vector Tobra viruses.

Tobacco rattle virus (genus *Tobravirus*) is a multi-component virus with rod-shaped long (RNA-1) and short (RNA-2).particles.

- M-type (normal) viral isolates contain both RNA-1 and RNA-2; they code for coat protein synthesis and are readily transmitted mechanically and by trichodorid nematodes.
- NM-type viral isolates have only contain RNA-1. They can be sap-transmitted and move through the plant systemically. Since they do not code for coat protein synthesis are not transmitted by trichodorid nematodes.
- The multi-component nature of the particles results in considerable variation among isolates.

Corky Ringspot of Potato (Spraing)

Corky ringspot is caused by tobacco rattle virus (Tobravirus), which is vectored by stubby-root nematodes (*Paratrichodorus* spp. and *Trichodorus* spp.). The condition is sometimes referred to as Spraing, a Scottish word meaning a bright streak or stripe.

Corky ringspot symptoms vary depending on virus strain, potato cultivar, and time of infection. Symptoms often include of brown necrotic rings, arcs, and diffuse spots which are considered quality defects and may result in after-harvest devaluation or rejection of either table or processed potatoes.

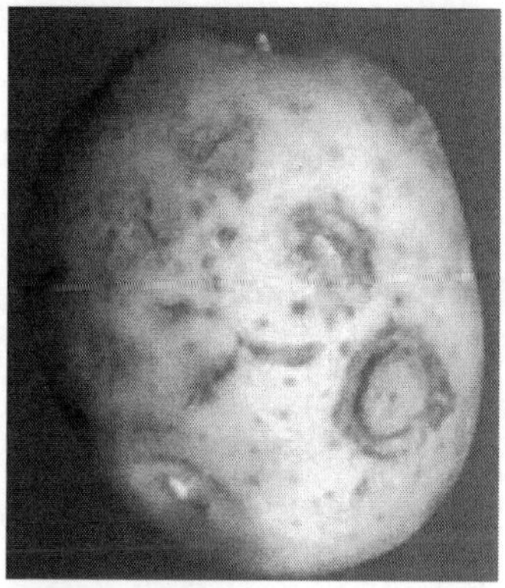

***Figure:** External Symptoms of Corky Ringspot Virus on a Potato Tuber*

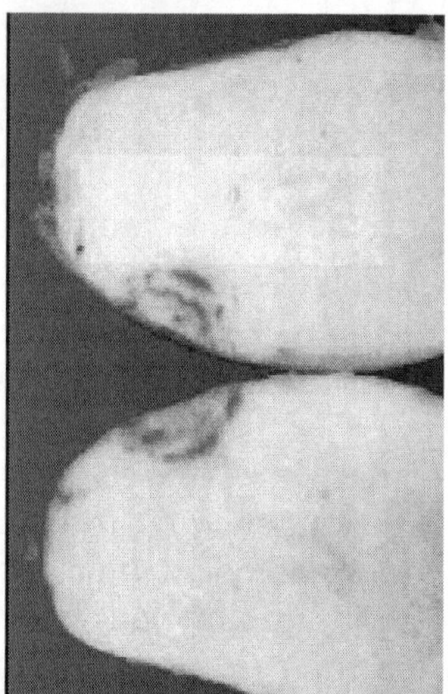

***Figure:** Internal Symptoms of Corky Ringspot Virus in a Potato Tuber*

Corky ringspot symptoms in potato include necrotic rings and pits on the tuber surface and range from diffuse brown spots to concentric rings or arcs of brown, necrotic tissue to dark-brown necrotic tissue which extends through tuber flesh. The virus is usually detectable when symptoms are seen but may also be present in asymptomatic tissue. Tubers from soil with a history of tobacco rattle virus serve as a reservoir. The virus may spread to daughter tubers when infected tubers are used as seed. Newly formed potato tubers are quite vulnerable to tobacco rattle virus infection and tubers as small as 3-cm in diameter had corky ringspot blemishes in tobacco-rattle infested fields in Florida. Tobacco Rattle Virus is reported in many areas of the world and more than 400 plant species in 50 plant families are susceptible to infection.

Several species of *Trichodorus* and *Paratrichodorus* transmit M-type isolates of Tobacco Rattle Virus. Stubby-root nematodes are migratory ectoparasites that are mobile during each stage of their life cycle and feed primarily on meristematic cells or root tips which hinders root elongation. Damaged root tips may swell and lateral roots may emerge behind them, resulting in root proliferation. The life cycle of stubby-root nematodes is not well studied. Eggs of *Paratrichodorus minor*, hatch 53.3 ± 7.3 degree-days using a basal threshold of 10° C (DD10C) after egg deposition. Second stage population peaks occurred at 28 DD10C, third stage at 67 DD10C, fourth stage at 109 DD10C and adults at 143 DD10C for adults. Population densities increased in the presence of a suitable host but declined to 33% of the initial level after 300 DD10C in the absence of a host. The optimum temperature range for development and reproduction of *Paratrichodorus allius* is 21 to 24° C.

In general, Trichodoridae nematodes prefer coarser textured soils and avoid desiccation stress by migrating vertically to moister soil layers. Vertical movement of Trichodoridae nematodes appears greatest when soil pores are half-full of water and least in waterlogged or dry soil conditions. Since soil moisture is often maintained below field capacity for optimum yield and quality in potato production, vertical migration may explain the difficulty of consistently finding stubby-root nematodes during routine soil sampling procedures to 30-cm depth in Washington and Oregon. Stubby-root feeding injury is not important economically and is rarely visible in potato, however, corky ringspot virus can be transmitted at very low population densities of the nematode.

Management Guidelines for Corky Ringspot Virus

The wide host range for both virus and vectors make eradication of corky ringspot in potato fields difficult. Alfalfa

(*Medicago sativa*), a rotation crop in Oregon and Washington, and spearmint (*Mentha cardiaca*) do not appear to serve as hosts of tobacco rattle virus. Since the virus is lost each time the nematode molts, and since it does not pass through the nematode egg stage, *P. allius* are cleansed of tobacco rattle virus after being reared on these plants for a period of 1 to 3 months. Therefore, it may be possible to eliminate tobacco rattle virus from populations of *P. allius* rotation crops that do not harbour the virus. Of course, the crops must be kept free of weed hosts for tobacco rattle for the strategy to be effective

Chemical Control of the Nematode Vector

Aldicarb (Temik 15G), a systemic carbamate pesticide, suppresses corky ringspot (damage in potato by controlling stubby-root nematodes.

1,3-dichloropropene was effective in reducing corky ringspot in Washington but not in Florida or the Klamath Basin of Oregon. The fumigation failure in Florida may be attributed to the ability of *P. minor* to escape the toxicant in deep soil and then migrate to upper levels soil profile to upper, treated areas once the active ingredient had dissipated.

The Role of Nematodes in Soil Fertility

Nematodes respond rapidly to disturbance and enrichment of their environment; increased microbial activity in soil leads to changes in the proportion of opportunistic bacterial feeders in a community. Over time the enrichment opportunists are followed by more general opportunists which include fungal feeders and different genera of bacterial feeders. This succession of nematode species plays a significant role in decomposition of soil organic matter, mineralization of plant nutrients and nutrient cycling.

Bacterial-feeding nematodes have a higher carbon:nitrogen (C:N) ratio (± 5.9) than their substrate (± 4.1) 12 (Ferris et al., 1997), so that in consuming bacteria they take in more N than necessary for their body structure. The excess nitrogen is excreted as ammonia. The C:N ratio of fungal-feeding nematodes is closer to that of their food source (Chen and Ferris, in press). However, for nematodes of both feeding habits, a considerable proportion of the C consumed is used in respiration (perhaps

40 % of the food intake. The N associated with respired C that is in excess of structural needs is also excreted. The excreted N is available in the soil solution for uptake by plants and by microbes. Because microbivorous nematodes exhibit a wide range of metabolic rates and behavioural attributes, the contribution of individual species to nitrogen cycling and soil fertility may vary considerably.

By developing C and N budgets for individual nematode species, we calculated that the bacterial-feeding nematode community in the top 15 cm of a field soil mineralized N at rates increasing to 1.01 µg-N g-soil-1 d-1 in the rhizosphere (Ferris et al., 1995, 1996, 1997). In microcosm experiments we determined that rates of N mineralization by bacterial-feeding nematode species of different body size ranged between 0.0012 and 0.0058 m g-N nematode^{-1} d^{-1}, mainly as NH_4^+. The amount of N as NH_4^+ released from microcosms in which barley straw was colonized by the fungus *Rhizoctonia* sp. was always greater in the presence of the fungal-feeding nematode *Aphelenchus avenae*.

In three years of field experiments we have demonstrated that N-availability in organic and low-input tomato production is enhanced by an abundance of microbial-feeding nematodes. The challenge is to manage the system so that those nematodes are in abundance at the start of the growing season. In our agricultural system we have shown that the constraining factor is usually soil moisture availability in the late summer and early fall of the previous year. By irrigating the soil at that time we can increase microbial activity and decomposition of crop residues. That provides substrate for the bacterial-feeding nematodes so that their population levels increase and remain at high levels until the next spring. In some cases the amount of C available to fuel the microbial community is also limiting and it may be necessary to incorporate organic material. Necessary approaches will vary with cropping system and climatic zone, but management of soil to enhance activity of bacterial- and fungal-feeding nematodes can increase soil fertility.

Soil nematode communities may also provide useful indicators of soil condition. Nematodes vary in sensitivity to pollutants and environmental disturbance. Recent development of indices that integrate the responses of different taxa and trophic groups to perturbation provides a powerful basis for analysis of faunal assemblages in soil as *in situ* environmental assessment systems. Application of nematode faunal composition analysis provides information on succession and changes in decomposition pathways in the soil food-web, nutrient status and soil fertility, acidity, and the effects of soil contaminants.

Nematode Community Dynamics

Dynamics of nematode communities in tomatoes grown in conventional and organic farming systems, and their impact on soil fertility.

A major management concern in low-input and organic farming systems is the maintenance of soil fertility. Nitrogen is supplied through crop residues, incorporated cover crops and manures in low-input and organic farming systems; it is applied to the soil in the conventional farming system as mineral N. Large amounts of carbon are included in the material required to achieve adequate N-levels in organic farming systems, so that they are carbon-rich while conventional systems are carbon-starved as interpreted on the basis of microbial biomass-C, microbial biomass-N, and measures of microbial activity.

Nematodes may play a significant role in decomposition of soil organic matter and mineralization of plant nutrients. Many of the nematodes in biologically active and productive soils are not herbivores but are bacterial-and fungal-feeding species. Primary decomposition is accomplished by bacteria and fungi, which are grazed upon by Protista and microbivorous nematodes. There is substantial evidence of the importance of the nematodes in mineral and nutrient cycling. Since bacterial-feeding nematodes are purported to have a higher C:N ratio (* 10:1) than their substrate (* 5:1), considerable mineralization of N is associated with their metabolic activity. In eating sufficient bacteria to provide the carbon necessary for their body structure and respiration, nematodes will take in more nitrogen than necessary. The excess nitrogen is excreted as ammonia. Even if the reported C:N ratios of nematodes are incorrect and their body composition is closer to that of bacteria, as suggested by analyses of Persson (1983), the nitrogen consumed with carbon that is used in respiration (perhaps 40 % of the food intake, will be in excess of structural needs and will be excreted. However, microbivorous nematodes exhibit a wide range of metabolic rates and behavioural attributes. The contribution of individual species to nitrogen cycling and soil fertility may vary considerably.

Soil fertility, microbial biomass and bacterial-feeding nematodes In a field experiment at UC Davis in 1993, the available nitrate in the soil solution differed during the growing season in the conventional and organic tomato plots. In the conventional system there was abundant inorganic N detected in soil samples between April and mid-June, resulting from preplant and sidedress applications of mineral fertilizer. Those levels were probably in excess of plant needs, but available during key growth periods for the tomato plants. In the organic plots, the level of inorganic N was very low at tomato planting and remained low throughout the growing season. It was below the norms for conventional agriculture and, under those criteria, the soil would be diagnosed as nitrogen deficient for tomato production. Plants exhibited nitrogen stress in the organic tomatoes during the early part of the growing season. However, crop yields in the two systems were comparable in 1993.

Microbial biomass, measured as MBC, was significantly higher in the organic farming system than the conventional system throughout the tomato growing season. Following cover crop incorporation in spring in the organic farming system, MBC increased until the end of June. After mid-summer MBC decreased and declined to levels comparable with those in the conventional plots by the end of the growing season. In the conventional plots, MBC remained relatively constant throughout the growing season.

Total numbers of bacterial-feeding nematodes were similar in conventional and organic farming systems at the time of tomato planting. It was more than 30 days after cover crop incorporation before the bacterial-feeding nematodes in the organic plots became more abundant than those in the conventional system. Numbers in the conventional system remained low throughout the growing season, while those in the organic system declined from their high levels following decrease in microbial biomass at mid-summer. By the end of the growing season the numbers in the two farming systems were not significantly different.

Nematode Species Composition and Abundance

Eleven taxa of bacterial-feeding nematodes, all in the order Rhabditida, were monitored during the course of this intensive sampling study. Those identified routinely to species were *Acrobeloides bodenheimeri* Thorne, *A. buetschlii* Steiner and Buhrer, *A. tricornis* Thorne, *Acrobeles* sp., *Cephalobus persegnis* Bastian and *Chiloplacus* sp. (all family Cephalobidae); *Bursilla labiata* Andrássy, *Cruznema tripartitum* Zullini, *Diploscapter* sp. and *Rhabditis cucumeris* Andrássy (all family Rhabditidae); and *Panagrolaimus detritophagus* Fuchs (family Panagrolaimidae).

Different taxa of bacterial-feeding nematodes predominated at different times during the growing season. In the organic plots, *Panagrolaimus* sp. and *Rhabditis* sp. were most abundant in the top 15 cm of soil early in the tomato growing season and within about 20 days of organic matter incorporation. *Bursilla* sp. became numerically predominant by the end of June in the organic plots, and remained at high population levels until the end of the tomato growing season. *Chiloplacus* sp. were more numerous in the conventional than the organic plots until near the end of the growing season.

Diversity Indices for Bacterial-feeding Nematodes

The bacterial-feeding nematode community in the conventionally-farmed plots was more diverse than that in the organic plots through most of the growing season. That reflected the temporal predominance of individual opportunistic species in the organic plots in response to the incorporation of organic matter and the subsequent flush of bacteria.

After incorporation of vetch at the beginning of the growing season, Bongers' Maturity Index suggested a higher level of maturity, that is a trend towards persister (K-selected) species, across all nematode species and trophic groups in the conventional than in the organic farming system. Conversely, there was a preponderance of opportunistic colonizer species in the organic system. As with other indices, the data are strongly influenced by the diversity among the bacterial-feeding nematode species in the two farming systems. Actually, the maturity indices for the bacterial-feeding nematodes are based on only two maturity groups.

Nematodes in the families Rhabditidae (*Bursilla* sp., *Cruznema* sp. and *Rhabditis* sp.) and Panagrolaimidae (*Panagrolaimus* sp.) predominated in the organic system. These nematodes are all in maturity group 1, the most opportunistic colonizers. In the conventional system nematodes in the family Cephalobidae (*Acrobeloides* spp., *Acrobeles* sp., *Cephalobus* sp. and *Chiloplacus* sp.) were more abundant. These nematodes are all in maturity group 2 and slightly less dynamic in their colonizing capabilities.

Biomass

Bursilla labiata, a small nematode abundant for much of the growing season in the organic farming system, had a maximum calculated biomass of 13.9 mg/l soil on June 21 at mid-season, many times larger than that of any other bacterial-feeding nematode in either farming system. Maximum biomass for all bacterial-feeding nematode species in both farming systems occurred at the June 21 sampling when it was 3.25 times greater in the organic (17.4 mg/l soil) than in the conventional (5.4 mg/l soil) system. At that date, the total number of bacterial-feeding nematodes counted in the organic system was only 2.25 times greater than in the conventional system. Although nematode numbers and biomass were similar at the time of tomato planting in both farming systems, both measures had become significantly greater in the organic system by May 10 and remained that way throughout the growing season.

Bacterial-feeding nematode biomass per unit of MBC during the first half of the growing season increased in the organic farming system, indicating that food availability was not a limiting constraint during that period. After the mid-summer decline in MBC there was a decline in nematode biomass per unit of MBC. That suggested that the bacterial-feeding nematode community was then greater than the carrying capacity of the environment and declined correspondingly. In the conventional farming system the biomass per unit of MBC remained relatively constant throughout the growing season, suggesting that the two measures were in dynamic equilibrium and that food supply was a constraint in nematode population and biomass increase in that system.

Discussion

Although population densities of bacterial-feeding nematodes at the beginning of April (day 95) were similar in the conventional and organic plots, individual species did not increase opportunistically in the conventional system as in the carbon-fed organic system. In the conventional systems, the organic matter incorporated into the soil is crop residue high in carbon content. The manure and leguminous cover crops incorporated into the low-input and organic plots have lower C:N ratios than the crop residues. It is possible that complex, high C:N ratio organic materials may select for fungal-rather than bacterial-dominated decomposition pathways. This may explain the greater numbers of fungal-feeding nematodes in the conventional plots early in the growing season. Similar observation have been made in the spring in other conventional and organic farming system comparisons and in differentiating between perennial and annual cropping systems.

Although the concept is the subject of ecological debate, the notion that diversity is an indicator of stability may provide some insights into the condition and resilience of soil systems. Diversity indices within a trophic group are of greatest interest since they reflect the level of redundancy of common function in the system. They are more likely to be true indicators of stability within a feeding category than indices constructed from unrelated organisms across trophic and functional levels. At any point in time there was less redundancy in common function among the bacterial-feeding nematodes of the organic farming system, perhaps suggesting that this component of the system would be readily perturbed by small disturbances.

Bongers' Maturity Index applied to bacterial-feeding nematodes provided an intriguing hint that the abundance of individuals in maturity group 1 (families Rhabditidae, Panagrolaimidae, Diplogasteridae and others) may be indicators of a biologically active soil with enhanced rates of N-mineralization. These nematodes may have similar metabolic rates as species in the Cephalobidae, but the dynamics of their population increase allow them to become more abundant in a shorter period of time. We infer that a large biomass of bacterial-feeding nematodes feeding on bacteria, there will be more N mineralized than by a smaller biomass with similar metabolic rates. We further infer that at this field site, the 13.9 mg/l soil of *Bursilla labiata* biomass, of a total bacterial-feeding nematode biomass of 17.4 mg/l soil in the organic system, indicate that this nematode is the most significant contributor to N-mineralization. We conclude that correction for extraction efficiency and calculation of biomass are important components of the determination of relative contribution of various species.

Clearly bacterial-feeding nematodes are more abundant, represent greater biomass, and consume greater numbers of bacteria in the organic farming system than in the conventional farming system by about one month after tomatoes are planted. Soil nitrate levels throughout the growing season, and especially early, were low. They were never as great in the organic system as in the conventional system, particularly during key periods of nitrogen demand during vegetative growth and fruit set. Plants in the organic system displayed symptoms of nitrogen deficiency early in the growing season. If bacterial-feeding nematodes are important contributors to N-mineralization, that contribution is minimal during the first month of the growing season but may be more important during fruit set. It seems desirable to increase their abundance and activity in N-mineralization early in the growing season in organic plots to meet the needs of young seedlings. That may require altering community structure by introducing large-bodied, actively-feeding nematodes, or changing cultural operations to influence the abundance, biomass and activity of bacterial-feeding nematodes, early in the growing season.

Since metabolic rates of bacterial-feeding nematodes from this field site are at extremely low levels at the soil temperatures experienced during the winter months, and cumulative physiological time increases slowly early in the growing season, it appears that an the appropriate strategy may be to maximize nematode abundance at the end of the growing season.

The most appropriate strategy for increasing their springtime abundance is to ensure that there are large population levels in the soil the previous fall. At that time their food abundance may be a limiting factor, although decline in numbers may also be influenced by abiotic factors such as cessation of irrigation prior to harvest.

A manure application at termination of the previous summer crop in the organic farming system, followed by irrigation, may enhance bacterial abundance in the fall and so maximize nematode abundance at that time. If overwinter survival is primarily mediated by abiotic (non density-dependent factors), a greater abundance of nematodes in the fall should result in a greater abundance in the spring. That would enhance the potential contribution of bacterial-feeding nematodes to N-mineralization during a critical plant growth period in the first month of the next summer crop.

The dynamics are such that spring incorporation of a winter cover crop to increase MBC is followed by too great a lag period in increase of the bacterial-feeding nematode community for effective impact on N-mineralization. The bacterial-feeding nematodes do not increase in the rhizospheres of the winter cover-crop because the cover-crop is planted too late and/or sown and left to germinate with winter rain. Soil temperatures by that time are below a level conducive to rapid reproduction of the nematodes. Early (mid-September) germination of the cover crop would promote biological activity and nematode increase through about mid-October. Since activity in the rhizosphere will be proportional to root size, which is initially zero, it will be enhanced by manure incorporation in September (as early as possible) and by following incorporated crop residue or incorporated manure with an irrigation to stimulate biological activity.

Chapter 12

Plant Nematology

This short contribution introduces the major nematode pests in developing agricultural systems and discusses possible measures for their control.

Root-knot Nematodes (*Meloidogyne* Spp.)

Meloidogyne species constitute the major nematode problem in developing countries. The most common species is *M. incognita,* which causes considerable losses in many crops. Many vegetable crops, in several botanical families, such as Solanaceae, Cucurbitaceae, Leguminosae, Liliaceae, Chenopodiaceae, Compositae, Umbelliferae, Cruciferae and Malvaceae, suffer the greatest damage. But *M. incognita* is also highly pathogenic to some staple crops such as cereals, including rice, maize, potato, soybean, banana, plantain, sweet potato and yam; or industrial crops such as tobacco, coffee, sugar cane, sugar beet, cotton and black pepper. Often this species also causes economic damage to fruit crops such as guava, pineapple, papaya and grapes.

Meloidogyne javanica is also widespread and prevalent in warmer climates. It attacks almost all the plants affected by *M. incognita* except cotton and pepper, although in the case of pepper there are a few reports. Many vegetable crops are severely affected by *M. arenaria* which, however, causes the most significant losses on groundnuts, especially in West Africa. *Meloidogyne hapla* has a more restricted distribution, being more common in cool climates or on winter crops in the Mediterranean area. Major economic damage has been reported on carrots, potatoes, alfalfa, clover and strawberries. A few other species of root-knot nematodes have restricted areas of distribution and affect locally various crops to different extents. Among them is *M. artiellia,* pathogenic to legumes and cereals in southern Europe and in the Near East. *Meloidogyne naasi,* which is widespread in northern Europe, has been reported affecting cereals in South America. *Meloidogyne exigua* and *M. decalineata* attack coffee in South America and Africa, respectively. *Meloidogyne brevicauda* is pathogenic to tea planted at high elevations in Sri Lanka. *Meloidogyne graminicola* affects cereals in India.

Cyst Nematodes (Globodera spp., Heterodera spp. and Punctodera Spp.)

Less cosmopolitan, but as pathogenic as root-knot nematodes, are species of cyst-forming nematodes. Among them the most widespread are certainly *Globodera rostochiensis* and *G. pallida.* They are very common in Central and South America where they are thought to be indigenous, but their occurrence has now been reported also in hilly African and Asian countries. They cause considerable damage in the western countries of North Africa.

Heterodera avenae is a common species in temperate areas of India where it is the causal agent of a disease of wheat called molya. It also attacks barley. *Heterodera schachtii* attacks sugar beet in North Africa, the Near East and Mexico. *Heterodera sacchari* is reported to damage sugar cane in Africa and India and rice in Africa. *Heterodera oryzae* parasitizes rice and banana in Africa. *Heterodera cajani* is perhaps the most pathogenic species in India on various leguminous crops such as pigeon pea, cowpea and chickpea. *Heterodera zeae* is very widespread and affects maize, sugar cane and sorghum from North Africa to India.

Of local importance, although destructive, are *H. daverti* and/or *H. trifolii* in North Africa. *H. ciceri* and *H. latipons* are other important nematode pests in the Near East, *and Punctodera chalcoensis* in Mexico.

Other Important Nematode Pests

Anguina tritici is a common parasite of wheat on the plateaus of Africa and Asia. In Asia and, to a lesser extent, in

Africa, *Aphelenchoides besseyi,* the causal agent of the white tip disease of rice, is responsible for major losses of this cereal. *Aphelenchoides ritzemabosi* is widespread in hilly areas of northern India on zinnia and *A. agarici* seems to be a destructive pest of cultivated mushrooms in India.

The stem nematode, *Ditylenchus dipsaci,* attacks several crops in the Mediterranean region and causes major losses of onion, garlic, broad bean and strawberry crops. *D. dipsaci* is also a major pest of garlic in many countries in South America. The phytopathological importance of *D. angustus* to rice in the Far East is also well known.

Within the genus *Helicotylenchus, H. dihystera* has worldwide distribution, being a pest of many crops. Its damaging effect, however, has been demonstrated only on olive in Egypt and on rice in Mauritius. *Helicotylenchus indicus* causes some damage to vegetable crops in India, but only when at high population densities and *H. multicinctus* occurs wherever bananas are cultivated. Fortunately, this species is a weak parasite affecting only the cortical tissues and causing economic damage only at high population densities. *Mesocriconema onoense, M. curvatum* and *Hemicycliophora typica* are parasites of rice in Africa.

Various species of *Pratylenchus* parasitize crops in developing countries. *Pratylenchus brachyurus* attacks many plants in Africa and in the Mediterranean region. Under tropical climatic conditions *P. coffeae* is an important pest of coffee, banana, cowpea and citrus. *Pratylenchus penetrans* is frequently associated with declining date-palms in Saharan oases. *Pratylenchus thornei* and *P. mediterraneus* are parasites of legumes and cereals in North Africa and the Near East and. *P. loosi* is a major pest of tea in Sri Lanka and India. Finally, *P. zeae* attacks maize and sugar cane in Africa and Asia.

Radopholus similis is a destructive parasite of banana throughout the tropics. It also causes great damage to black pepper in Java, to coconut palms in India and to tea in Sri Lanka.

Species of *Hirschmanniella* such as *H. oryzae, H. gracilis, H. spinicauda* and *H. mexicana* cause varying extents of damage to rice in Africa, Asia and Mexico.

Hoplolaimus seinhorsti has been shown to be pathogenic to black pepper and various vegetable crops in Sri Lanka, and *H. indicus* and *H. pararobustus* are often found associated with a decline of sugar cane and other plants in Africa and Asia. Yams in Africa are severely affected by *Scutellonema bradys,* and it is also an important postharvest pest attacking yams in storehouses.

Hemicriconemoides mangiferae may be the most widely associated nematode species with mango. It occurs mainly in Asia and Africa where it is found also in the rhizosphere of other fruit-trees. Among *Tylenchorhynchus* species, *T. brassicae* is the most frequently occurring in the rhizosphere of crops. It is associated with different plants in Africa and Asia. Other common species are *T. claytoni* and *T. indicus.*

Rotylenchulus reniformis is widespread in all the regions with tropical and subtropical agriculture. It causes economic damage to many plants, including vegetable crops, fruit-trees and staple and fibre crops. The citrus nematode, *Tylenchulus semipenetrans* is invariably associated with citrus groves, which are affected to different extents depending on the paedoclimatic and edaphic situations. *Rhadinaphelenchus cocophilus,* the causal agent of red ring disease, is very destructive on coconut and oil-palms in Central and South America.

There are species of *Longidorus, Paralongidorus* and *Xiphinema* that may cause serious problems, although restricted to local situations. *Longidorus africanus* is a parasite of cotton in Mexico and the Sudan, but it also occurs in the Near East and East Africa. *Longidorus laevicapitatus* attacks sugar cane in many African countries, *Longidorus cohni* is a pest to cereals and alfalfa in the Near East, *Paralongidorus buchae* attacks many crops in Mauritius, *P. oryzae* is a parasite of rice in India and *P. citri* has been consistently found associated with declining groundnut crops in India.

Among the *Xiphinema* species, the most common in the tropics, from West Africa to India and Sri Lanka, are *X. ifacolum,* which is highly pathogenic to rice and many vegetable crops, and *X. elongatum* which is a parasite of olive in Egypt and sugar cane in West Africa and India.

Methods of Control

Possible control measures vary with the paedoclimatic conditions, the socioeconomic situation, the economics of the crop, the availability of chemicals and/or resistant cultivars, and the feasibility of some agricultural practices.

Crop rotations can always be adopted against species with narrow host ranges. However, plant sequence and time intervals between susceptible crops depend on the nematode species. For ectoparasites, with long life cycles and with a few generations each growing season, one-or two-year intervals might be appropriate. In the case of root-knot nematodes, or at least with species that lay eggs in a gelatinous matrix, at least three years between susceptible crops must elapse. But when cyst nematodes occur, even five years may not be sufficient to obtain economic results. High-value cash crops are often grown intensively by specialist growers and so it may be difficult to find a suitable plant sequence that will not alter the economic and cultural balance of a farm.

Resistant cultivars are without doubt the easiest and most convenient choice for the farmer. However, they are not always available, they may bring on to the market innovations not easily and quickly accepted by consumers, and they can develop resistant pathotypes or races of the parasite if used in monoculture for many years. Nematicidal treatments have been successfully used, but nematicides are usually expensive and may raise problems of environmental pollution and/or of accumulation of toxic residues in edible plant products. Methyl bromide is one of the most effective fumigants to control a broad range of problems (nematodes, fungi, insects, weeds). Being highly toxic to humans and highly volatile, it must be applied by skilled personnel and under a plastic tarp. Removal of tarping should be done two days after treatment and planting not earlier than one week after application. The high cost of the operation, lack of licensed operators and unavailability of the product, which requires proper storage facilities, are limiting factors for the use of this chemical.

A cheaper fumigant with predominant nematicidal effect is 1,3 dichloropropene (1,3-D). Its volatility is much lower compared with methyl bromide, therefore it does not require tarping. However, to be effective it needs to be injected into the soil by efficient machines which are not always available in developing countries and require accurate maintenance practices. The easiest and most effective way of applying 1,3-D is by injector gun, but this requires much labour. An interval of at least three weeks is necessary between application of this chemical and planting to avoid phytotoxicity.

When fumigants are applied, the soil must be carefully prepared by breaking plough pans and large soil aggregates and removing residues of the previous crop. Soil humidity and temperature must also be optimal to achieve satisfactory results.

Granular nematicides are more easily applied and safer for farmers compared with fumigants. However, they need high soil humidity to be effective. Those with systemic action are usually very effective and persistent.

Aldicarb, a carbamate, exerts its nematicidal activity for over six months, but it may cause problems of residues in edible plants. Therefore its use is allowed mainly on ornamental and industrial crops. Fenamiphos persists in the soil for about two months and if applied at rates of 15 to 20 kg a.i./ha gives good control even of root-knot and cyst nematodes without inducing accumulation of toxic residues in plants. It should be applied preferably two to three weeks before planting. Oxamyl is also a systemic nematicide but has shorter persistence (two weeks) and is less effective in tropical climates than carbofuran. Prophos performs its activity only by contact. Therefore it is effective against ectoparasitic nematodes and can control endoparasites only before they penetrate the roots. The limiting factor of all these chemicals is their high cost which makes them affordable only for cash crops. Even then, they are often not available on the market because hard currency is needed to import them, which is usually scarce in developing countries.

Two other pesticides that have mainly fungicidal activity have been used to control nematodes. To be effective dazomet needs moist soil and an interval of about six weeks between treatment and planting because of its high phytotoxicity. Metam-sodium has the same mode of action as dazomet, but has the advantage of being applicable through the irrigation system a few weeks before planting, and the disadvantage of having only modest nematicidal activity. Therefore, it needs very high rates of application to control root-knot or cyst nematodes, which are in most circumstances the only plant nematodes whose control gives economic benefit.

The hypothesis has been advanced that continuous application of nematicides might develop resistance to them in some nematodes. Up to now there is no evidence of resistance of nematodes to chemicals, and the cases in which low mortality of nematodes has been observed following nematicidal treatments have been interpreted as an enhanced degradation of the chemical itself resulting from the increased activity of microorganisms.

Attempts have been made to control plant-parasitic nematodes by soil amendments of by-products of some agricultural industries, such as neem, mustard, castor bean, groundnut or coconut oilcakes. In fact they reduce numbers of nematodes and female fecundity, but they are effective only when modest populations occur.

Plant extracts such as those from *Tagetes, Ruta, Cineraria or Pelargonium* have also been effective in killing plant-parasitic nematodes, but results refer mainly to *in vitro* or pot experiments and practical application of these extracts has yet to be profitable.

Soil steaming is often suggested as an environmentally clean measure to control plant pathogens. However, this time-consuming method can be applied only on small areas and, for the amount of energy (oil or electric power) required and the sophisticated equipment needed, is one of the most expensive nematode control practices. It is worth while only in greenhouses for production of out-of-season luxury foods. Besides, it does not always give uniform results and, if excessive temperatures are produced, reduces soil fertility and deteriorates soil structure.

Soil solarization is a recently developed technique in warm areas. It is a hydrothermal process that occurs in moist soil covered with transparent plastic film in summer months. During solarization, soil temperatures reach values (42° to 50°C) that are lethal to plant-parasitic nematodes, or increase to 36° to 40°C which, if prolonged, may be lethal or may affect life cycles and host-parasite relationships of many plant-parasitic nematodes by depleting their energy reserves. This makes them more vulnerable to biotic (antagonistic organisms) and abiotic stresses. However, temperatures of solarized soil are seldom lethal or sublethal at levels deeper than 30 cm. Therefore it is most practical with annual crops with shallow root systems. Its effect can be further improved if used in combination with reduced rates of fumigants or non-volatile nematicides.

Biological control may be feasible if antagonistic organisms are cheap to mass produce and apply. Easy colonization of soil, high virulence, long persistence, good shelf-life, low sensitivity towards agrochemicals, safety for operators and the environment and effective nematode control at acceptable costs for farmers are important factors. Unfortunately the fact that none of the organisms so far used for biological control of nematodes possesses all these characteristics and the lack of technology for producing commercial preparations of such biocontrol agents are the main limitations to this practice.

In conclusion, different methods of nematode control could be combined in an integrated system, shifting from the concept of control to the concept of management which is a procedure directed at reducing and maintaining the numbers of plant-parasitic nematodes at non-injurious levels. This is particularly important in developing countries, where the lack of a commercial system and of a market for wealthy customers often means that high production costs are not justified. However, demographic increases require improvements in production from sustainable agriculture.

Techniques and Methodologies for Nematode Disease Diagnosis and Nematode Identification

Most plant-parasitic nematodes occur in soil around roots and are ectoparasitic, but many endoparasitic species are found abundantly in rhizosphere soil. Some plant-parasitic species are not important economically since they do not cause significant damage to plants. When they do cause noticeable damage they are considered pathogenic. Soil-inhabiting ectoparasitic forms become important when their population inflates to reach the so-called economic threshold. For any control programme, whether chemical, biological, physical or cultural, accurate nematode identification is of the utmost importance. Even determining pathotype or race may become important in a particular group, e.g. Heteroderidae. It is also important to know the hosts of a species and how important it is pathologically to plant growth and yield.

Major nematode parasites are polyphagous and invade many plant hosts, multiply quickly to have several generations per year and have easy means for spread and dispersal. Examples are seen in many root parasites and some parasites of above-ground plant parts (*Aphelenchoides ritzemabosi, A. besseyi, Ditylenchus dipsaci*). Seed-borne pathogenic nematodes (*Anguina tritici, Aphelenchoides arachidis, Ditylenchus dipsaci*) have developed special adaptive mechanisms, such as the ability to withstand desiccation, for their survival and dispersal. Species like *Rhadinaphelenchus cocophilus, Bursaphelenchus xylophilus, Deladenus* spp., *Fergusobia* spp., use insects as vectors/hosts. Host specificity is shown only by some groups of nematodes, e.g. Heteroderidae, Meloidogynidae, Pratylenchidae, Neotylenchidae. Such host-specific nematodes are supposed to have co-evolved along with their hosts. Co-evolution and zoo-geographical distribution of plant nematodes can throw some light on their origin, evolution and radiation.

Besides their normal mode of transmission, parasites, pests and pathogens are consistently being moved about by people. Human activities and cargo traffic by land, sea and air have tremendously increased the chances of dispersal of plant nematodes. Exotic pests are constantly being introduced into clean areas. The nematodes are brought in with exotic plants, imported grain, vegetable and fruit produce and in soil attached to packing material, machinery, tyres of motor vehicles, tools and even shoes. Live third-stage juveniles, which is the dispersal stage of the pinewood nematode *Bursaphelenchus xylophilus,* were recovered in Finland from pine boards used in a packing case imported from Canada. Dispersal through wind, rain and floodwater respects no national boundaries.

The potato cyst nematode (*Heterodera rostochiensis*) possibly originated in the Andean region of South America and was imported into the United Kingdom during the Irish famine of 1845/46 on potatoes brought for resistance against *Phytophthora infestans,* the causal agent of the Irish famine.

From there it was distributed to mainland Europe, India and Australia. Similarly, *Heterodera avenae* on wheat was distributed by people. Many temperate nematodes are flourishing in high altitudes in the tropics because they were transported there with planting material. Application of detection and identification techniques are basic to checking for plant nematode introduction, spread and distribution.

Knowledge of hosts and geographical distribution of important species is essential. For example, false root-knot nematode, *Nacobbus* spp. and *Monotrichodorus* spp., occur only in the New World, although the former may occur under glass in Europe. *Heterodera zeae* was found on *Zea mays* and described in India. Later it was discovered in Maryland, United States, on an indigenous monocot by a stream. It was immediately put under quarantine and it was thought that it had come from India. Since it was found on an indigenous plant in the United States, it must be indigenous to that country. Very likely it originated in Mexico and co-evolved with maize and other monocots and was distributed with maize by humans.

Surveying, Collecting, Fixing and Mounting Nematodes

Surveying large areas for the presence or absence of plant nematodes is important but difficult. Systematic surveys are conducted using statistical designs to obtain a reliable estimate of nematode abundance. Fields and patches in a large area should be selected at random and sampling procedure should be based on standardized sampling techniques. Nematode population assays should be made to determine both the qualitative and quantitative occurrences of important species and to relate their population levels to crop damage. Non-agricultural areas adjoining cultivated land and non-crop plants in and outside that area should also be surveyed to identify potential pest problems. This will help an understanding of the alternative hosts and aspects of the biology of nematode species. Biological data including damage symptoms, life cycles, population trends and agroclimatic information on soil types, climatic conditions, cropping sequences and the times of sampling are also important in surveys because populations are dependent on them. This knowledge is of utmost importance when control measures are contemplated.

Surveys should have backing from diagnostic and identification centres to obtain quick and reliable identifications. Fields and large areas for surveys must be selected at random and statistical sampling designs must be used to obtain quick and reliable data on nematode prevalence. Periodic large-scale surveys of important species are essential to assess and monitor nematode populations. They should be repeated every two to four years to determine the relative abundance of species and new introductions, and there should be a system of coordinating various surveys to have a clear picture of the host and distribution of important species within an area. Coordination of surveys and pooling of data at an appropriate centre for analyses are important. About 250 to 300 ml of soil from around roots is collected in a polythene bag and data on host, locality, soil type, etc. are tagged. Care should be taken to ensure that soil samples are moist during transit. About 100 g of roots including sound, partially attacked and well-attacked parts, and a similar quantity of shoots showing apparent damage, are collected and stored in polythene bags. Soil from roots of adjoining grasses and weeds should not be allowed to get mixed with the samples. Auger samplers are used to obtain soil cores for population studies at different depths. Care must be taken that soil is not mixed with soil from above or below the segment to be analysed.

Samples should be processed soon after collection. They may be stored at about 4°C for later extraction. Extraction methods must be directed towards isolating all stages of nematode development. Variously modified Baermann techniques and the Cobb sieving and gravity method are used for obtaining quick and reliable results. Their aims are to recover actively migrating nematodes, and their drawbacks include missing out sedentary nematodes. Many nematodes lie dormant in the soil or plant tissue and need soaking in water for several hours before extraction. A simple bucket-sieving method for isolating nematodes from soil samples of about 300 ml, using a large bucket and a 45 to 53m m aperture sieve, is described in Siddiqi (1986). The nematodes thus extracted are separated from debris and mineral particles by pouring the suspension on to a wire mesh covered with tissue paper and submerged in water in a trough. The active nematodes pass through the filter paper into clean water and can be concentrated by decantation or centrifugation. Soil and plant material can be treated similarly to extract live, motile nematodes. The centrifugal sugar-flotation technique and elutriators are good for processing several samples and obtaining all stages of active or inactive nematodes.

Nematodes may be killed by pouring hot water (85° to 95 °C) over them. In doing so, nematodes should be in the smallest possible quantity of water so that they are exposed to instant heat. They may be fixed in 3 to 5 percent formaldehyde solution and preserved in the same medium. Nematode structures are best studied when examined in water, fixative or lactophenol.

For making permanent mounts, the nematodes are transferred to glycerine and mounted in the same medium. Several methods are described in Siddiqi (1986) for making permanent nematode slides. Siddiqi (1964) described a quick method of transferring nematodes to glycerine by processing through warm lactophenol. The nematodes are transferred to warm lactophenol held in a cavity slide or watch glass. They will first shrink but later stretch out to their normal shape within two to three minutes. If they do not do so and remain shrunk, then more heat should be applied until they are fully stretched. Then the slide can be left at room temperature overnight. The slide is again warmed and a few drops of a mixture of 75 percent glycerine and 25 percent lactophenol are poured on to the nematodes held in thick, concentrated lactophenol. The slide is kept warm for five to ten minutes. Finally the nematodes are transferred to glycerine. Sufficient time should be allowed for the glycerine to penetrate the nematodes, which then can be mounted in pure dehydrated glycerine.

For mounting, nematodes already in glycerine are transferred to a small drop of glycerine placed on a clean glass slide and surrounded by glass fibre pieces for support. Three small lumps of paraffin wax (melting point 55° to 60°C) are put around but separated from the glycerine and a 19 mm diameter cover slip is placed over the wax lumps. The slide is then heated until the wax melts and fills the space between the cover slip and the slide. Finally the slide is sealed with glyceel.

Detecting Nematode Diseases

Detection and identification of nematodes is the first step in controlling them and checking their spread. This may result in economic savings of great magnitude. For example, *Radopholus similis* was recognized as the causal agent of the spreading decline of citrus in Florida in 1953. Since then, regulatory agencies in neighbouring states have imposed stringent measures to control its entry and spread. A constant vigil has saved the citrus industry of other states of the United States from being devastated by this disease. *Heterodera glycine* of soybean (*Glycine max*), occurs in the Far East and United States. It has been detected and identified in Egypt. Unless strict quarantine measures are applied, the nematode will soon spread in the countries of the Near East.

Nematode diseases usually go unnoticed as infected plants are rarely killed. A particular crop may show only patches of damage where the nematode populations are much larger than in the rest of the field. At times, entire fields may be infested and overall health of the plants may appear poor. The yield from such infested plants is considerably reduced. Farmers usually blame unfavourable soil or climatic conditions, but if the nematode disease is controlled, for example by the application of nematicides, yield is dramatically increased.

General above-ground symptoms of nematode infection include unthrifty reduced growth, severe stunting, chlorosis and wilting of leaves and reduced quality and quantity of produce. Root damage includes swelling and knotting; shrivelling, splitting and cracking; brown necrotic lesions; and stubby, broken or girded rootlets.

Root symptoms are produced by three kinds of reaction:

- A hyperplastic reaction results in hypertrophy and hyperplasia in which both cell number and cell size are increased. Simple hyperplastic galling of root tips is caused by the feeding of *Xiphinema, Hemicycliophora* and *Belonolaimus* spp. Feeding by *Meloidogyne* spp. results in severe hypertrophy and hyperplasia of cells; multinucleate giant cells are formed at the feeding site in the cortex, endodermis and pericycle.

- A hypoplastic reaction that inhibits growth in meristematic tissue and causes stunted and stubby roots is caused by the feeding of tylenchorhynchids and trichodorids, respectively.
- A necrotic reaction is caused by nematode feeding, their movement within the tissues and the presence of microorganisms. Browning and shrivelling of feeder roots is caused by ectoparasitic nematodes. *Pratylenchus* spp. and *Radopholus* spp. cause extensive necrotic lesions in roots by their feeding and movement. The necrotic reaction is caused by enzyme secretion into plant tissue and tissue response, as well as by the accumulation of phenolic compounds in the infected area of the root. Secondary invasion by bacteria and fungi aggravates damage.

Diseases may be caused by individual nematode species, or by a combination of several species or nematodes interacting with other pathogens to produce disease complexes. The disease complexes produced by the interaction of nematodes with pathogenic bacteria or fungi are more damaging to plants than these pathogens acting alone. Some nematodes are vectors of plant viruses. *Xiphinema index* was first shown to be able to transmit the fanleaf virus from diseased to healthy grape-vines. Now we know of several longidorids and trichodorids which transmit the soilborne viruses of plants.

Nematodes have evolved various types of association with plants. The parasitic ones may be ecto-or endoparasites; they may be migratory or sedentary parasites. Many are host-specific but most are polyphagous. Siddiqi (1983) gave a detailed analysis of the ecological adaptations of plant-parasitic nematodes.

Migratory Parasites of Above-ground Plant Parts

Aphelenchoides besseyi, the causal agent of white tip disease of rice, is a migratory ectoparasite and is seed-borne. Infested seeds, when soaked in water for 24 hours, yield motile nematodes. The nematode also attacks above-ground parts of strawberries causing summer dwarf or crimp disease.

The bud and leaf nematodes (*Aphelenchoides ritzemabosi* and *A. fragariae*) feed ectoparasitically on buds and parenchymatous cells of the leaf mesophyll. *A. ritzemabosi* parasitizes leaves of chrysanthemum and other ornamentals. The symptoms produced include spots or blotches on leaves, distorted tissue and stunted and deformed buds. Soaking the infested plant tissues in water releases the live nematodes.

Red ring disease of coconut is caused by *Rhadinaphelenchus cocophilus* in the Caribbean and northern parts of South America. It is transmitted to new trees by the palm weevil, *Rhynchophorus palmarum*. *Bursaphelenchus lignicolus* causes a destructive disease of pine trees in Japan. It infests axial and radial resin canals and is pathogenic causing the death of pine trees. It is transmitted by the cerambycid beetle, *Monochasmus alternatus*.

Stem and bulb nematode *Ditylenchus dipsaci* was first described on teasel heads in Europe. It attacks stems of pea, beans, alfalfa etc. and bulbs such as onion (causing bloat disease), daffodil and narcissus. In the Sudan and the Syrian Arab Republic, the infested bean and onion fields exhibit patchy areas and larger populations occur in wet rather than in dry areas.

Heavily attacked onion and narcissus bulbs show rings of decayed tissue when they are cut. When infested stem or bulb tissues are macerated in water, hundreds of nematodes in various stages of development are liberated.

In a diseased bulb, stem or seed, heavily infested with *D. dipsaci*, the pre-adult juveniles aggregate in thousands and roll up to form what is known as nematode wool which enables them to survive desiccation and other unfavourable conditions.

Ditylenchus angustus feeds ectoparasitically on rice stems, leaves and inflorescences producing various types of malformation of tissue. It causes Ufra disease of rice in Bangladesh and is known to occur in Myanmar, India and Madagascar. *D. drepanocercus* infests leaves of an evergreen tree, *Evodia roxburghiana*, in India.

Sedentary Parasites of Above-ground Plant Parts

Members of the family Anguinidae (*Subanguina, Anguina, Cynipanguina, Nothanguina, Orrina, Pterotylenchus*) produce galls on above-ground parts of plants; only one species of anguinids, *Subanguina radicicola,* produces galls on roots. *Anguina tritici* and *A. agrostis* form seed galls on wheat and grasses, respectively. *A. tritici* causes ear cockle disease of wheat resulting in wheat grains becoming deformed, brown or black galls that transmit the nematodes when sown along with healthy grains. Tundu disease of wheat is caused by the interaction of *A. tritici* with *Corynebacterium tritici.*

Anguina funesta produces seed galls on rye grass in Australia and acts as a vector of the toxin-producing bacterium, *Corynebacterium rathayi.* About 1 000 bacteria-infested galls are lethal to a sheep. Leaf galls are produced by *Subanguina, Cynipanguina, Nothanguina, Orrina* and stem galls are produced by *Ditylenchus, Pterotylenchus* etc. *Orrina phyllobia* is a parasite of silverleaf nightshade (*Solanum elaeagnifolium*) and can be used as a biocontrol agent for this perennial weed that infests cotton fields in southwestern United States.

Fergusobia spp., of the family Neotylenchidae, produce aerial shoot galls on *Eucalyptus* spp. in Australia and *Syzygium cumini* in India. They show dual parasitism of trees and agromyzid flies (*Fergusonina* spp.). Teasing out galls in water releases all stages of nematode development.

Migratory Ectoparasites of Roots

Most plant-parasitic nematode species are ectoparasitic. All plant-parasitic members of Dorylaimida, Triplonchida and Criconematina are ectoparasitic. In Criconematina, the male has a degenerate stylet and oesophagus and cannot feed. Migratory ectoparasitic nematodes may have small stylets (Tylenchidae, *Merlinius*) to feed on root hairs and epidermal cells, or long stylets (*Hemicycliophora, Amplimerlinius*) to feed on deeper tissues. Long and strong stylet-bearing *Hoplolaimus galeatus* and *Belonolaimus* spp. are pests of lawn and turf grasses in Florida. They produce yellow patches in which the root system is considerably reduced in size as a result of the feeding of nematodes. The ectoparasites, *Xiphinema* and *Tylenchorhynchus,* kill epidermal cells by feeding and cause discoloration and superficial necrosis. Stubby root and docking disorder in crops are caused by trichodorid nematodes.

Species of *Xiphinema* and *Longidorus* are known to transmit soilborne NEPO viruses of plants. *Xiphinema index* and *X. diversicaudatum* transmit fanleaf virus of grapevines and arabis mosaic virus, respectively. *Longidorus elongatus* is a vector of raspberry ringspot virus and tobacco black ring virus. *Trichodorus* and *Paratrichodorus* transmit TOBRA viruses of plants. Plants showing virus disease symptoms should be sampled for these nematodes.

The ectoparasitic nematodes are generally polyphagous and survive in soil without a host for many years. They are easily extracted from soil by Baermann funnel, sieving or flotation techniques.

Sedentary Ectoparasites of Underground Plant Parts

Sedentary obese females of the ectoparasitic genera *Tylenchulus, Sphaeronema, Rotylenchulus,* etc. are detected only by examining the roots. The author's report of the finding of *Tylenchulus semipenetrans* and *Rotylenchulus reniformis* in India in 1961 was the result of collecting the motile stages of these species by a sieving method using 250-mesh sieves. The immature female stage and the male of the latter species gave rise to the idea of creating a new genus for them since they were unique to Hoplolaimidae. The author had not seen the kidney-shaped mature female, but the timely warning of Dr M.W. Alien of California University prevented a synonym of *Rotylenchulus* being proposed, as authors of *Leiperotylenchus* and *Spyrotylenchus* had done before. For identification of these nematodes, therefore, all stages of development should be examined.

Tylenchulus semipenetrans causes slow decline of citrus and occurs in all areas of citrus culture. The female's anterior end lies deep in the root reaching the stele region where it feeds, while its body, most of which remains outside the root, swells. The root tissues respond to nematode feeding by producing nurse cells near the head of the nematode, thus avoiding destruction and decay. The nematodes may aggregate and form colonies on roots and lay eggs in a protective gelatinous material. Juveniles of *T. semipenetrans* remain viable in soil for several years after the citrus trees are removed. Recovery of these juveniles from the rhizosphere soil of plants that might have grown there is therefore not a proof of such plants being the hosts of this nematode.

Migratory Endoparasites of Underground Plant Parts

Pratylenchus and *Radopholus* spp. live, feed and multiply within the root and underground stem tissue. *Pratylenchus* spp. cause necrotic lesions in the epidermis and cortical parenchyma that may extend to the endodermis and stele. The lesions eventually girdle the rootlets and interrupt the water and nutrient supply of infected plants. Often such lesions are invaded by rot-causing bacteria and fungi, completely destroying the rootlets. *Radopholus similis similis* is a serious pest of banana, coconut, tea and black pepper. It causes severe root rot that, in the case of banana, results in black head or toppling disease, and in black pepper the yellows disease. *R. similis citrophilus* causes the spreading decline of citrus in Florida. The diseased trees show wilting of leaves, dieback of twigs and general unthriftiness, low fruit yield and a poor root system in which the feeder roots are shortened and reduced to less than half the normal size and number.

Scutellonema bradys infests yam tubers and even multiplies in stored yams causing severe damage. *Hirschmanniella oryzae, H. mucronata* and *H. spinicauda* parasitize rice roots in many tropical and subtropical countries. The nematodes feed and move about intercellularly in air channels between the radial lamellae and the parenchyma of the roots. *Hirschmanniella miticausa* parasitizes taro (*Colocasia esculenta*) causing mitimiti disease in the southern highlands of Papua New Guinea. *Aphelenchoides arachidis* infests groundnut pod shells and seed testa in northern Nigeria, is resistant to desiccation and is seed-borne.

Evolutionary Concept and Phylogenetic Relationships

Taxonomy is the science of identifying and naming animals and plants and assigning them to groups in a system of classification. The identification and naming of individual taxon (plural taxa) is called microtaxonomy, while placing them in a hierarchical system of classification is macrotaxonomy. Taxa include any category from subspecies to superfamily. Taxa higher than superfamily, e.g. order or class, are not covered by the *International Code of Zoological Nomenclature.* Taxa are usually diagnosed on the basis of phenotypic similarity/

dissimilarity and not on shared common ancestry (as they should be).

Macrotaxonomy as a science of classification is based on the philosophy of organic evolution. All animals and plants are continuously changing, as are all non-living things in the universe. Some changes are brought about in a short time and are observable, but others take thousands of years and are not seen or sensed. Macrotaxonomy uses various methodologies to understand the mechanism and degree of this long-term change. These methodologies utilize the characters/features of animals/ plants to determine their primitive or derived nature by assessing their weighting and homology. Conclusions are then drawn about their relationships for building a hierarchical system of classification.

Relationships are interpreted in terms of descent from a common ancestor. This is not an easy task since interpreting homology increases the amount of speculation. Several approaches have been developed to establish the relationships of taxa and to classify them. Siddiqi (1986) in his book on Tylenchida has discussed at length cladistic, evolutionary and phenetic approaches. However, in classification, many nematologists are not using these approaches at all, but are largely depending on their own concepts and experiences of the groups. It is said that identification work, which is different from phylogenetic work, should not necessarily depend on such approaches, but new species and genera, when proposed, should always be properly diagnosed and the criteria for their differentiation be discussed to show their relationships with existing taxa.

Morpho-anatomical Methods

Species identification utilizes characters/features of many kinds including morpho-anatomical, physiological, ecological, ethological, embryological and cytogenetical. By far the most useful and widely used characters are the morpho-anatomical.

A comparative study of morpho-anatomical characters/ features in a range of closely related species should be made to evaluate their importance in identification. This would include the study of each character's stability/variability within that group. Hasan (1990) states: "Many species occur in mixture in natural populations. In most cases, due to lack of data on interbreeding, reproductive biology, physiology, ecology and geographical distribution etc., the concept of *phenetic species* has been employed in nematode taxonomy for diagnosing a taxon. Such a concept has its own importance, however many of these species have been known to exhibit a wide range of intraspecific variations in morphology and dimensions which according to animal taxonomists is the function of geographical isolation, ecological/environmental stresses and genetic inequality."

The biospecies concept involves determining a species on the basis of whether populations interbreed if their members are morphologically similar to each other but different from other such populations. Difficulties arise when either the members of the populations are morphologically similar but do not interbreed, or they do interbreed but are morphologically dissimilar. *Ditylenchus dipsaci* populations from phlox and others do not interbreed successfully. Are they then separate species? *Globodera rostochiensis* and *G. pallida* differ morphologically, but they may mate freely in laboratory experiments and are expected to do so in the field. A female could mate with several males and thus the offspring in a cyst may not represent a pure culture.

Interbreeding as a test for species recognition does not apply to parthenogenetic forms, and among plant-parasitic nematodes such forms are quite common. Morpho-anatomical determination of species is therefore of utmost importance in parthenogenetic forms and is the main method of identification. Its application to sibling species, however, becomes difficult. For example, *Radopholus similis* and *R. citrophilus* are similar in morpho-anatomical details but not so in physiological and cytogenetic characters. In such sibling species, the concept of subspecies is worth while. The two species may be called *Radopholus similis similis* and *R. similis citrophilus* to show their relatedness but to keep them as distinct forms. The occurrence of closely related species sympatrically which maintain their morphological identity gives them a good taxonomic status.

Parthenogenetic species are prone to clone-formation because they lack bisexuality and hence the exchange of genetic material. Such clones differ in small morphological differences and thus naming them as separate species could be dangerous. Variations among large populations from different places should therefore be studied thoroughly. These variations are due to isolation, geographical separation, host effects and several physical and chemical factors. Species will, however, continue to be described on morpho-anatomical characters.

The superspecies concept, although not used in plant nematology, is a good one and includes, under one name, morphologically closely related species, sibling or cryptic species and subspecies, e.g. *Globodera tabacum* can be a superspecies to include *G. tabacum, G. virginiae* and *G. solanacearum.*

Morphometric and Morpho-anatomical Characters

Since plant nematodes are microscopic, morpho-anatomical characters have to be correctly observed and interpreted. Even taxonomists fail to observe some characters correctly. For example, some *Helicotylenchus* spp. show a distinct dorsal oesophageal gland but indistinct subventral glands and so they have been described under *Rotylenchulus/Orientylus.* A solid-appearing anterior region of the stylet conus *of Tylenchorhynchus* and a conus having an asymmetrical lumen in *Histotylenchus* can only be studied properly when seen in lateral view. A post-anal intestinal sac was wrongly reported in *Tylenchorhynchus stegus* and *Quinisulcius himalayae* possibly because the granules in the lateral hypodermal chords looked similar to those in the intestine. Phasmids were reported in the tail region of the Tylenchidae and of *Ditylenchus* although they are absent in these groups.

Detailed studies on morpho-anatomical characters including their variability in many populations from many hosts and geographic regions provide the basis for creating new species. Taylor and Jenkins (1957) described *Pratylenchus hexincisus* sp.n. and *P. subpenetrans* sp. n. and Fortuner (1970) described *Aphelenchoides siddiqii* sp. n. because they studied a number of closely related populations. The same applies to the author's description of new species of *Monotrichodorus, viz. M. acuparvus, M. parvus* and *M. proporifer* from South America.

Morphometric data. Morphometric characters including the de Manian ratios should not be the sole basis for species differentiation because they are often very variable. Taylor and Jenkins (1957) found L, a, b, c, c' ratios *in Pratylenchus* spp. as highly variable, but the V ratio showed a greater stability. The V ratio is a good taxonomic character which is used to differentiate species of many genera, e.g. *Pratylenchus, Paratylenchus, Rotylenchulus.* In several cases, overlapping ranges of many morphometric characters make it very difficult to use them for species differentiation.

Sex and sexual dimorphism. Presence/absence of males and of sexual dimorphism are used as differentiating features. Where males are absent, the female spermatheca is reduced and empty. Sexual dimorphism in body shape is a good character of Heteroderidae and Rotylenchulinae and in the anterior region (cephalic region, stylet and oesophagus) is a characteristic of the Radopholinae as against the Pratylenchinae.

Body size and shape. Body length, width and shape (cylindrical, tapering at ends, kidney-shaped, lemon-shaped etc.) and habitus at death (straight, arcuate, spiral etc.) serve as useful differentiating characters. It should be kept in mind that fixation and processing to glycerine may cause reductions, distortions and artefacts. Therefore, specimens in water or fixative should always be measured and studied.

Cuticle. Cuticle thickness, striation, annulation (both transverse and longitudinal), punctation and ornamentation, and cuticular modification (ridges, spines, scales, alae) serve as useful characters as do the lateral field and their incisures. Thickening of the cuticle at the tail tip is a good character for *Trophurus* and *Paratrophurus.* Longitudinal ridges characterize *Mulkorhynchus* while the presence of five or six lateral incisures are used to diagnose *Quinisulcius* and the Merliniinae, respectively. Body pores are used as taxonomic characters in Trichodoridae and the excretory pore position is used as a taxonomic character. Amphids, deirids, phasmids and sensory papillae including male supplementary papillae are very useful in the taxonomy of all groups. Bursal ribs on the tail are diagnostic for the Aphelenchidae while the presence of hypoptygma around the cloacal aperture differentiates the Merliniinae from the Telotylenchinae.

Cephalic region. The shape and the degree of separation from the body of the cephalic region is used in species differentiation. Its transverse and longitudinal striation, the formation of a labial disc and the degree of sclerotization of its framework are also important.

Stylet and oesophagus. By far the most diagnostic characters are found in the stylet and oesophagus. Length of the stylet and the relative length of its conus are useful, as are the size and shape of the stylet knobs. Siddiqi (1971, 1986) discussed the importance of the oesophagus in the classification of the Tylenchida. Among the Tylenchina, the families Tylenchidae, Psilenchidae and Dolichodoridae have a basal bulb enclosing the three oesophageal glands, while in the Hoplolaimidae, Heteroderidae and Meloidogynidae the glands lie free in the body cavity extending over the intestine. In the Hoplolaimidae, the glands may extend mostly ventrally or dorsally, but in the Heteroderidae and the Meloidogynidae, they are always ventral to the intestine. The dorsal gland nucleus is larger than those of the subventrals in the Xiphinematinae, while it is smaller than the subventrals in the Longidorinae.

The presence of the orifice of the dorsal oesophageal gland within the median bulb is one of the diagnostic characters of the order Aphelenchida. In the same group, the presence of a distinct isthmus is the characteristic feature of the Aphelenchoidea. The shape and size of the cardia or oesophago-intestinal junction are useful in the Dorylaimida.

Intestine, prerectum, rectum and anus. The number of cells in a cross-section of intestine and the length of the prerectum are useful in the taxonomy of the Dorylaimida. The anus in members of the Aphelenchida is a large, posteriorly directed crescentic slit compared with the small and pore-like structure in members of the Tylenchida. Rectum length in respect to anal body width is also used as a character. A post-anal intestinal sac differentiates between *Bitylenchus* and *Tylenchorhynchus* and its length is useful in differentiating species.

Female reproductive system. Reproductive systems can be didelphic, monodelphic, pseudomonodelphic (when one branch is secondarily reduced), prodelphic, opisthodelphic, monoprodelphic and mono-opisthodelphic. The length of the reproductive branch as a percentage of the body length is important in some cases, but in others it is highly variable since it depends on the degree of development of the ovary. The shape and location of the vulva is a good character, as is the presence/absence of lateral vulval membranes and epiptygma. The shape and size of the spermatheca and its position in respect to the genital branch (axial or offset) are good diagnostic characters. When a branch is reduced to a sac, its length and the presence/absence of a reduced ovary serve as good characters, e.g. in *Pratylenchus*. The length of the mature ovary and the arrangement of oocytes are also important, e.g. in *Aphelenchoides*.

Male reproductive system. The size and shape of the spicules, gubernaculum and spermatozoa are useful characters. There is a distinct difference in the shapes of the spicula and gubernacula of the Tylenchorhynchinae versus the Merliniinae and the gubernacula of the Pratylenchinae versus Radopholinae. The shape of spermatozoa differs in species of *Radopholus*. Genital papillae and ventromedian supplements are important characters of trichodorid and dorylaimid nematodes. Copulatory muscles were used by Siddiqi (1974) to differentiate between *Trichodorus* and *Paratrichodorus* and the presence of hypoptygma was used by Siddiqi (1970) to differentiate *Merlinius* from *Tylenchorhynchus*.

Tail. Tail length and shape are used in taxonomy but they show considerable variation. The shape of the tail tip, although a useful character in many groups, was shown to be variable in some species *of Pratylenchus*. Nevertheless, a pattern of tail tip shape emerges when a large number of specimens are studied. For example, *Pratylenchus panamaensis, P. coffeae* and *P. loosi* have characteristic tail tip shapes. The tail is consistently elongated in the Tylenchidae, hooked in the Halenchidae and short and rounded in females of the Hoplolaiminae and males of the Heteroderidae and the Meloidogynidae.

Juvenile characters. Tail shapes in juveniles of various stages is a good character in the differentiation of *Xiphinema* species. Presence of spined juveniles in *Hemicriconemoides* differentiates it from *Hemicycliophora*. The shape and arrangement of spines in juveniles of this genus are used for species differentiation.

Biological, Biochemical and Cytogenetic Methods

Biological characters, especially host preference, can be used in the identification of some plant-parasitic nematodes.

However, the effects of physical, chemical and biological factors on host-parasite relationships are often great. Presence or absence of males is also a differentiating character, but it is known that at least in some species, males may arise as a response to environmental stress. Biochemical and cytogenetic techniques are used to discover the degree of genetic similarity in a taxonomic group. Such data must be obtained for several species to compare with both in-group and out-group members. A comparative study of such data provides valuable information about characters/features that are the result of common genetic material, as well as those that are unique for a particular member of the group. The uniqueness determines the identity of the taxon. Genetically dependent molecular data can be used concordantly with the morpho-anatomical characters to determine natural groupings and evolutionary trends.

Isozyme phenotypes, particularly esterases, have been used in the identification of *Meloidogyne* spp.. An inventory of esterase/phenotypes is made for a comparative study. Although having great potential, polyclonal (PCA) and monoclonal (MCA) antibodies have been utilized so far only in a limited area for the detection and identification of nematode pests. Immunofluorescence and antisera could provide valuable tools for the detection of nematode diseases and identifying and classifying plant-parasitic nematodes. Biochemical characterization makes identification simple and quick.

Serological techniques using PCA and MCA have been tried and some specific antigens from endoparasitic nematodes have been identified and used, for example in separating *Globodera pallida* from *G. rostochiensis.* Davies and Lander (1992) found a large number of MCA using three *Meloidogyne* spp., *M. incognita, M. javanica* and *M. arenaria,* but none was specific and each cross-reacted with at least one other species. However, three MCA were promising in distinguishing between the three species on ELISA-based and immunoblotting assays.

Cytogenetic techniques to determine chromosome number have been used successfully in the classification of higher taxa and the identification of species and pathotypes of *Meloidogyne* and Pratylenchidae. All the Criconematoidea have a basic chromosome number n=5. Dolichodoroidea have n=8 and Heteroderidae and Meloidogynidae have n=9 (polyploidy 18) and n=18 (polyploidy 36, 54), respectively. *Radopholus similis citrophilus* was differentiated from 7?. *similis similis* in having n=5 compared with n=4 in the latter. *Meloidogyne* species can be differentiated by determining chromosome number, differential host tests and studying perineal patterns.

Nematode Identification Expertise, Equipment and Aids, Literature and Nematode Reference Collections

Nematode Taxonomists

Experts are needed for the development and implementation of any scientific programme. Nematode taxonomists are few and far between and they have been branded as a dying breed. The importance of plant-parasitic nematodes in agriculture, the role of nematodes in ecosystems and the potential use and need for the preservation of biological diversity in non-agricultural lands, have increased the demand for nematode taxonomists. However, the demand cannot be met unless taxonomic teaching and training programmes are implemented and funds to support and sustain them are found.

In the Near East region, there are good taxonomists and diagnostic centres only in Egypt, the Islamic Republic of Iran and Pakistan. But even in these countries, nematode species are not properly surveyed and catalogued. Further surveys and taxonomic research programmes should therefore be launched. For this, support funds should be obtained from national and international funding agencies and collaborative work between the taxonomists of the region should be encouraged.

Microscopes and Computers

For any diagnostic work, a good compound microscope is essential. Fitting interference equipment to the stereoscope compound microscope provides sharp images and reduces eye strain. The use of the scanning electron microscope (SEM) has increased in the study of nematode species. New species descriptions often include SEM photomicrographs that show clearly such characters as the amphids, labial and cephalic papillae, lateral field incisures and genital papillae. However, for routine identification work, SEM is not essential. Diagnostic characters, if based only on SEM observations, will be difficult to use. Nevertheless, *Radopholus similis similis* and *R. similis citrophilus,* which are not differentiated by morpho-anatomical characters studied with a light microscope, are readily differentiated by examining the SEM photomicrographs of the cloacal region. Computers are now being used for recording and analysing identification data. Survey records of parasites, hosts, occurrences, soil types, climatic conditions etc. become so large that manual sorting and analysis take an immense amount of time and labour. Also, it is difficult to edit and update the records. All such records and data can now be stored in computerized filing systems providing databases that can be searched, retrieved and sorted within a few minutes. It also avoids storage problems and the possibility of accidental damage. Databases can easily be copied into duplicate diskettes to avoid accidental loss.

Need for a joint effort in computerizing survey records and developing computer-aided identification

In the Near East region, there is hardly any centre where survey records and nematode collections are computerized. Computers are rarely used in identification work. Esser (1991) produced a computer-ready check-list of plant-parasitic nematode genera and species to be used in identification work. Computer-based identification expert systems are now being developed. A joint effort of 33 taxonomists from 13 countries with support given by the NATO Science Programme and the North Carolina State University, reviewed identification methods and proposed the development and implementation of an expert system for plant-parasitic nematode identification. Countries in the Near East region should try to fund such a project in one of their countries to serve the needs of the entire region.

A system for networking all information on occurrences of plant-parasitic nematodes, including host lists in the Near East region, is a dire need on which due attention must be focused. Such information will be useful to all the countries in the region in checking the spread of pests, parasites and pathogens.

Literature

Availability of literature is the most important requirement of a diagnostic and identification centre. Many countries in the

Near East region do not even have sufficient funds to subscribe to the core nematological journals, with the result that the papers produced by their workers are often poor in quality. Monographs, checklists, dichotomous and polytomous keys, taxonomic data sheets etc. are useful tools that make identification work comparatively easy. Esser (1991) used Siddiqi's (1986) book on the Tylenchida extensively in his computerized checklist and remarked: "After 1974, M.R. Siddiqi's 1986 Tyienchida is a very valuable source of data". Esser produced many identification aids as keys and compendia which are widely used as identification tools.

Nematode Type and Reference Collections

The importance of type and reference collections in nematode identification cannot be exaggerated. Identification studies invariably involve comparing test animals with specimens of other species. Basing comparisons on published data alone is dangerous since many species descriptions and even illustrations can be poor and sometimes misleading.

High quality and reliable identifications are subject to good taxonomic expertise, and comparison with actual specimens of related species is essential. Taxonomists should always deposit the type specimens of their new species in recognized nematode collections where they can be properly looked after and made available to other workers, if required. Good type and reference collections are available at the national depositories in India (IARI, New Delhi) and Pakistan (NNRC, Karachi) and in international centres such as the International Institute of Parasitology, St Albans, and Rothamsted Experiment Station, United Kingdom.

Nematodes

Nematodes are invertebrate roundworms that inhabit marine, freshwater, and terrestrial environments. They comprise the phylum Nematoda (or Nemata) which includes parasites of plants and of animals, including humans, as well as species that feed on bacteria, fungi, algae, and on other nematodes. Four out of every five multicellular animals on the planet are nematodes. Cobb (1914) calculated that if the nematodes resident in a single acre of soil near San Antonio,

Texas, USA, were to proceed in head-to-tail procession to Washington D.C., some 2000 miles away, the first nematode would reach Washington before the rear of the procession left San Antonio!

The majority of nematodes are microscopic, averaging less than a millimetre in length, but some of the animal parasites are quite large and readily visible to the naked eye. The animal and plant parasites are of direct importance in agriculture, the environment, and in human health; however, most nematodes in the environment are not parasites. Nematodes that feed on other organisms are important participants in the cycling of minerals and nutrients in the ecosystem that is fundamental to other biological activity. Some of these nematodes may have major roles in decomposition, including biodegradation of toxic compounds. In fact, the incidence of certain nematode species is sometimes used as an indicator of environmental quality. Insect-parasitic nematodes can be of importance in regulating insect populations, and are being used in the biological control of insect pests.

The developmental biology of one nematode species, *Caenorhabditis elegans*, is better characterized than that of any other multicellular organism. *C. elegans* is studied as a model system in molecular and developmental biology, and is providing insights into many other areas of biology and medicine.

Agriculture

The management of plant-parasitic nematodes has been fundamental to advances in agricultural production in the United States and worldwide. However, the use of certain pesticides, such as 1,2-Dibromo-3-chloropropane (DBCP) to control nematodes, has resulted in contamination of soil and groundwater in California. Current research seeks integrated and sustainable approaches to the management of nematode populations through genetic manipulation of host plants, design of cropping systems, development of suppressive chemicals of botanical origin, and biological control. There is currently a major effort to shift from nematicide-based nematode management to alternative methods. Some 15 million pounds of 1,3-Dichloropropene have been used annually on approximately 200,000 acres in California.

Alternatives to this approach are critical, particularly following the April 13, 1990 suspension by the California Department of Food and Agriculture (CDFA) of the use of 1,3-Dichloropropene. Since alternative approaches are more target-specific and system-specific, priorities will have to be established for research and development in applied nematology. At the same time, strong linkages to more basic areas of nematode biology must be fostered and maintained to promote the development of new technology.

Animals

Of the nineteen Orders in the phylum Nematoda, seven contain nematodes that are parasites or associates of invertebrates, and six include species that are parasites of vertebrate animals. Nematodes are reported as parasites and associates of many invertebrate animals, especially in the Annelida, Mollusca, and Arthropoda.

In some cases, the invertebrate functions as the intermediate host in a life-cycle that includes parasitism of a vertebrate. In other cases, the invertebrate, usually an insect, functions as a vector between vertebrate hosts, or the nematode is passively transported by the insect. Several interesting plant-parasitic nematodes fall into this latter group and, significantly, they are closely related to nematode species that are parasites of insects. A considerable research effort has been applied toward using nematode parasites of insects as biological control agents, e.g., for mosquitos and blackflies.

Some of the nematode associates of insects are important because they vector bacteria that kill the insect. The nematode invades (or is consumed by) the insect, and bacteria are released into the insect hemolymph. When the insect is dead or near death, growth and subsequent development of nematodes occur as they utilize essential steroids supplied by the insect. These nematodes are also used extensively in the biological control of insects and are particularly effective against those insects that pass through at least one life stage in the soil.

Some 5,000 species of nematodes are estimated to be parasites of vertebrate animals and humans. These species are often characterized in a larger group of worm parasites as helminths. Nematode parasites of domestic vertebrate animals are managed by strategies that include control of secondary hosts or vectors and the use of chemical anthelminthics. Helminth infections of wild animals are, of course, not managed, except by attrition of infected individuals.

As human demography patterns change in California, and throughout the world, the interface between the ranges and habitats of wild and domestic animals change and overlap. Consequently, the pattern of exposure of domestic animals to helminth infections is also changing, and new associations continue to be reported; for example, the incidence of heartworm (*Dirofilaria immitis*) infection in dogs is currently increasing in California.

In general, the nematode parasites of California wildlife have only been studied descriptively. There is much interesting biology to investigate. Studies on the diplotriaenid nematode parasites of the air sacs of California swallows indicate that the birds carry a substantial biomass of nematode parasites on their annual migrations. The studies raise interesting ecological questions regarding flight efficiency and energetics, and also provide models for considering the distribution of parasites.

Both freshwater and marine fish are subject to nematode infections. The impact of the infections on the health and longevity of fish in nature is generally unknown. Frequently, nematodes are observed in the tissues of fish purchased by consumers.

The nematodes are usually killed during cooking, but certainly the transfer of live fish parasites to humans can occur during consumption of sashimi and other raw fish products. Generally, these nematodes will not establish a permanent infection in humans, but they may cause intestinal disorders in attempting to do so.

Humans

There are other well-known examples of the transfer of nematodes to humans. In most cases, the incidence of infection is relatively low due to regulatory inspection of food products, public education, and cooking of food. An example is trichinosis caused by the nematode *Trichinella spiralis*. Humans become infected by *Trichinella* by eating raw or undercooked pork.

The nematode parasites of humans cause a variety of disease conditions and symptoms, ranging from lack of energy and vigor to blindness and malformations. Pinworms, hookworms, and roundworms are extremely common intestinal helminth infections of humans; worldwide, roundworms are probably the most common, but in the U.S., pinworms predominate. Pinworm transmittal generally occurs through ingestion of fecal-contaminated material, and infection occurs commonly in children.

Other helminth infections are vectored as filarial worms by insects such as mosquitos, or the filaria may penetrate directly through the skin from water or soil. Filarial worms cause such diseases as river blindness (*Onchocerca volvulus*) and elephantiasis which are major health problems in some third-world countries. In the United States, most helminth infections of humans are controlled by public health programs, public education, vector control, intermediate host control, and anthelminthic drugs. However, changing demographic patterns, including the immigration of new California residents from third-world countries, has resulted in the introduction of unfamiliar helminth infections into the state. Frequently, the faculty in the Departments of Nematology are consulted by public health officials for identification of unfamiliar nematodes.

Forestry

In California, forestry is important for lumber, lumber products, and recreation. Assessment of the nematode impact on lumber yield and quality in forestry, however, is especially difficult since harvest cycles may be as long as 75 years, and land geography and soil composition are extremely diverse. Nematode impact may be via plant-parasitism; fungal feeding by nematodes, resulting in destruction of mycorrhizae essential to forest growth; and above-ground parasitic nematode parasites vectored by insects, e.g., pinewood nematode.

Preliminary investigations have indicated that there is an entirely new array of nematode pests indigenous to California forests, in addition to known agricultural pests, which feed on native forestry species. It is important that we begin to accumulate baseline information on nematode pests of California forests, particularly since no such programs exist in the western United States.

Recent plans for shipment of raw lumber from Siberia for processing in northern California may assuage public concerns regarding deforestation in the U.S.; however, analyses of Siberian logs have revealed a high potential for introduction of new pests. Included among the pests detected in the lumber is the nematode, *Bursaphelenchus mucronatus*, and its associated insect vectors. *B. mucronatus* is not reported from the U.S. and is a close relative of the pinewood nematode that has devastated forests in Japan.

Although the pinewood nematode already occurs in the U.S., its detection in wood chips exported to Scandinavia has resulted in a ban on those products. Certainly there are approaches to mitigating the potential for introduction of new pests, including debarking and fumigation of logs prior to shipment.

Rangeland

With respect to rangeland, it is necessary to consider a different array of nematodes than those which impact agriculture and forestry. The nematode problems in range include not only root and foliar parasites, but also diseases resulting from nematode-microorganismal associations. The latter aspect is perhaps best illustrated by a nematode-bacterial association in which the grass forage becomes toxic to livestock; this situation constitutes a major problem in Australia where large numbers of sheep and, in some areas, entire unattended flocks have been killed.

This nematode (*Anguina agrostis*) and the associated bacterium (*Clavibacter* sp.) also exist in northern California and southern Oregon. Because the livestock and dairy industries constitute a major component of agricultural production in California, and because rangeland and forests constitute approximately two-thirds of California land area, the significance of nematode problems on rangeland, and the need for additional research in this area, becomes clear.

Direct control techniques used for nematode management in traditional agriculture are most likely impractical for range situations. The resolution of range and forest problems depends upon alternative technologies which, in turn, depend on extensive baseline information incorporating nematode characteristics, plant characteristics, and environmental and edaphic factors.

Aquatic Systems

Nematodes are, by nature, aquatic organisms. It is estimated that about 50% of nematode species inhabit marine environments, although many of these have yet to be described and characterized. The remainder of the species inhabit soil and freshwater. In the soil, their aquatic requirements are satisfied by inhabiting the water films around soil particles. Parasitic nematodes are biologically active when bathed in moisture films supplied by water in the tissues or body fluids of the host.

Analyses of nematode communities in aquatic environments reveals that the incidence and prevalence of species in the community reflect the nature and quality of the environment. Not surprisingly, the types of species present (and the resultant community structure) differ in marine, brackish, and freshwater environments.

Recent observations indicate that various nematode species respond differently to degradation of environmental quality. Thus, the degree and nature of change in the community structure of aquatic nematodes may be an excellent indicator of water quality or pollutant levels. Nematodes in freshwater aquatic systems also serve as a nutrient source for invertebrates, small vertebrates, and fungi. The source of food for these nematodes is primarily bacteria, but algae and fungi are also consumed.

A considerable number of plant-parasitic nematodes in aquatic systems are associated with higher plants, although the impact of their parasitism on those plants is generally unknown. Preliminary research indicates a potential for management of these nematodes in the biological control of aquatic weeds. Genera that parasitize crop plants grown in immersed culture (e.g., rice) are well-characterized and are extremely important crop pests worldwide.

Marine Nematodes

The marine environment provides habitat for an enormous diversity of nematodes, from surface, littoral and estuarine zones to the ocean depths. One interesting group of deep sea nematodes are the Rhaptothyridae, which have no mouth and a very reduced alimentary tract. The digestive tract is filled with symbiotic chemoaototrophic bacteria. A similar relationship exists in the mouthless genus *Astomonema.*

Urban

We have only begun to understand the nature of the nematode problems and the biology of nematodes in the urban environment. As the demographics, economy, and land-use patterns in the state change, research in urban nematology will become increasingly important. Within the scope of urban nematology, are individual, societal, and commercial components.

Individuals create a personal environment involving trees, shrubs, other annual and perennial ornamentals, ground covers, lawns, indoor plants, and vegetable gardens. Frequently, the plants are placed directly into soil already containing plant-parasitic nematodes from a previous planting; nematodes may also be introduced with the current planting in associated soil and roots. Obviously, the layout and plant species involved in urban gardening are not selected for incompatibilities of their associated nematode species.

Nematodes introduced or supported by one plant species can be very damaging to neighbouring or companion plants. Such individual plants have aesthetic value or generate emotional attachments and, therefore, have a high individual value. Nematode damage to lawns and groundcover often results in poor growth, bare spots, and chlorotic appearance which the individual attempts to remedy by increasing the application of both fertilizer and water.

Societal components of urban nematology include nematode effects in parks, recreation areas, landscape plantings, highways, schools, etc. Nematode problems of turf constitute an area of concern for recreational and landscape industries.

Again, all these plantings have tremendous value in terms of management input and public perception. Damage and lack of plant vigor due to nematode parasitism result in unsightly areas, increased weed competition, and increased water and fertilizer usage. Commercial components of urban nematology include the costs of nematode management-or of lack of nematode management-by producers of plants in nurseries, turf farms, and greenhouses. The economics of urban nematology also include the activities of agrichemical industries, wholesalers, retailers, advisors and consultants.

As with other areas of nematology, we should consider the positive and negative aspects of nematodes in the urban environment. Nematodes can act primarily as pests or pathogens, or can also participate in decomposer food chains by using dead organisms as sources of carbon and energy. Often, the most productive soils contain high populations of non-parasitic nematodes participating in the cycling of minerals and nutrients. In fact, high population levels of such nematodes may be an indicator of productive soils with a healthy biological status.

As knowledge of nematode problems in urban situations increases, so will the potential to design spatial patterns and sequences of backyard gardens to provide the ultimate in nematode management options. Planting sequences and companion plantings can be selected with nematode susceptibilities, or even allelopathic effects, in mind. Enhancing the biological activity of soil to improve nutrient and moisture status, as well as the biological antagonism of nematodes, can be accomplished with minimal cost through the intensive efforts of avid gardeners in small-scale areas. The backyard garden offers the possibility of creating a microcosmic "sustainable agriculture" situation. Here is an environment that the homeowner can control and manipulate experimentally without the overriding concerns of marketability or profitability of crops.

Experimentation with soil manipulation, incorporation of organic matter, companion plants, sequences, and spatial patterns of plants may be a rich source of innovations and provide important biological insights. Biological antagonists of nematodes can be introduced and tested; beneficial nematodes for the biocontrol of insect pests can also be introduced and conserved in the system.

The opportunities for observation, formulation, and testing of hypotheses are more readily provided by the backyard garden than by commercial agriculture. To capitalize on those opportunities, many research questions must be addressed; these include plant species compatibilities (in terms of host status or effects on various nematode species), and knowledge of the nematodes present at a given site and their virulence on ornamental plants. Conceivably, increased regulation may be required for industries that impact urban nematology in order to minimize nematode introductions and spread in the urban environment.

Bibliography

Alfred Steferud: *Diseases of Fruits and Nuts*, Biotech Books, Delhi, 2005.

Bahar A. Siddiqui and Samiullah Khan: *Plant Breeding Advances and in vitro Culture*, CBS, Delhi, 1997.

Bandyopadhyay, P C : *Breeding and Crop Production*, Gene Tech Books, Delhi, 2007.

Boyer , J.S.: *Measuring the Water Status of Plants and Soils,* Academic Press, N.Y., 1995.

Coste R.: *Coffee: the Plant and the Product*, London, MacMillan, 1992.

Dharamvir Hota: *Modern Biotechnology in Plant Breeding*, Gene Tech Books, Delhi, 2007.

Dodia, DA ; IS Patel and GM Patel: *Botanical Pesticides for Pest Management*, Scientific, Delhi, 2008.

Fageria, N K ; V C Baligar and R B Clark: *Physiology of Crop Production*, International Book, Delhi, 2007.

Gupta, O.P. : *Water in Relation to Soils and Plants : With Special Reference to Agriculture*, Agrobios, Delhi, 2002.

Hartley W.: *A Checklist of Economic Plants in Australia*, C.S.I.R.O., Melbourne, 1979.

Herbert Meschkowski: *Ways of Thought of Great Mathematicians*, Holden-Day Inc., San Francisco, 1964.

Herklots G. A. C.: *Vegetables in South East Asia*, London, George Allen & Unwin Ltd., 1972.

Hessayon D. G. Dr.: *The Vegetable Expert*, PBI, Publications, England, 1985.

Huaman Z.: *Descriptors for Sweet Potato*, International Board for Plant Genetic Resources, Rome, 1991.

Jones, R. M.: *Plant Resources of South-East Asia,* Wageningen, Pudoc Scientific Publishers, 1992.

Kapoor, R.L. and M.L. Saini: *Plant Breeding and Crop Improvement*, CBS, Delhi, 1997.

Kataria, T N : *Plant and Crop Physiology*, Pearl Books, Delhi, 2008.

Lorenzen, S.: *The Phylogenetic Systematics of Freeliving Nematodes*, London, The Ray Society, 1994.

Mitra S.: *Postharvest Physiology and Storage of Tropical and Subtropical Fruits*, Oxon, CABI, 1997.

Nobel, P. S.: *Physicochemical and Environmental Plant Physiology*, Academic Press, San Diego, 1999.

Oldham P.: *Cost of Production of Major Tree Crops in Dominica*, Roseau, Ministry of Agriculture, 1991.

Parry M.L.: *Climatic Change, Agriculture and Settlements*, Dawson Folkestone UK, 1978.

Pemberton, R. W.: *Predictable Risk to Native Plants in Weed Biological Control,* Oecologia, 2000.

Pilgrim R.: *Post Harvest Handling of Minor Exotics*, St. George's, Grenada, 1996.

Politycka, B and C L Goswami: *Plant Physiology : Research Methods*, Scientific, Delhi, 2007.

Ramann, E. : *The Evolution and Classification of Soils*, Asiatic Pub, Calcutta, 2006.

Singh, S K : *Biotechnology, Plant Propagation and Plant Breeding*, Campus Books, Delhi, 2008.

Tripathi, Pramila : *Botanical Pesticides in the Management of Post-harvest Fruit Diseases*, Daya, Delhi, 2005.

Trivedi, P C : *Plant Physiology : Current Trends*, Pointer, Delhi, 2007.

Tyagi, I.D. : *Plant Breeding and Genetics at a Glance*, South Asian, Delhi, 2005.

Vanderplank, J.E.: *Plant Diseases: Epidemics and Control,* New York, NY, USA, Academic Press, 1963.

Verma, L.R. and R.C. Sharma: *Diseases of Horticultural Crops: Fruits*, Indus, Delhi, 1999.

Walia, S. and B.S. Parmar: *Pesticides, Crop Protection and Environment*, Oxford & IBH, Delhi, 1995.

Index